The Unseen Universe: Unveiling the Secrets of the Early Cosmos

Phillips

Contents

Chapter 1

Introduction

1.1 The early universe and heavy-ion collisions

The early universe was composed of a hot mixture of particles moving at nearly the speed of light [1]. This hot mixture was a result of the extremely high temperatures of the universe, which was too hot for any hadrons to have formed. As the universe expanded and cooled down, it experienced numerous phase transitions such as the electroweak phase transition where most of the known elementary particles acquired their Higgs masses. This phase transition took place at $\sim 10^{-11}$ s after the Big Bang when the temperatures of the universe were ~ 100 GeV ($\sim 10^{15}$ K) [2–4]. The electroweak phase transition was followed by the strong phase transition, where the quarks and gluons became confined into colour neutral hadrons and there was spontaneous chiral symmetry breaking [5]. This transition occurred at $\sim 10^{-5}$ s after the Big Bang at temperatures of ~ 200 MeV ($\sim 10^{12}$ K) and is understood to be a crossover from lattice calculations [6]. Another phase transition worth mentioning is the era of recombination which occurred approximately 10^{13} s after the Big Bang where the cosmic background radiation was released and the opaque plasma neutralised into a transparent gas [7, 8]. An approximate timeline of the universe is shown in Fig. (1.1).

Cosmological observations (i.e. through the use of telescopes) have been consistent with our understanding of the Big Bang nucleosynthesis from lattice QCD calculations [9]. However, these observations cannot be used to "see" directly the hot QCD matter that the universe was composed of a few microseconds after the Big Bang. Furthermore, we can't even cosmologically see the imprint of the transition from the hot quark matter to colour neutral hadrons [9]. Due to these limitations that cosmological observations have, to study and understand some properties of this hot quark matter, we turn to heavy-ion collisions.

The use of powerful particle accelerators has allowed us to recreate conditions of the early universe through ultrarelativistic heavy ion collisions. These accelerators include the Relativistic Heavy Ion Collider (RHIC) which has studied mostly $Au + Au$ collisions at $\sqrt{s_{NN}} = 200$ GeV

[10, 11] and the Large Hadron Collider (LHC) which studies $Pb+Pb$ collisions at higher energies (i.e. TeV scale) [12]. One can investigate whether these ultrarelativistic heavy ion collisions are actually producing the hot QCD matter that filled the universe a few microseconds after the Big Bang and that we are interested in studying. We can perform this investigation by looking at estimates of the properties of the matter that is produced in heavy-ion collisions and comparing these properties to those of normal nuclear matter.

The energy density produced in heavy-ion collision systems is very high [1, 9]. An estimate of the average energy density approximately 1 fm/c after the collision in a central $Pb + Pb$ collision with $|y| < 0.5$ at $\sqrt{s_{NN}} = 2.76$ TeV is shown in Eq. (1.1) [9, 13].

$$\epsilon = \frac{1}{(\pi R^2 \tau)} \times \frac{dE_T}{dy} = \frac{1.65 \text{ TeV}}{\pi (7 \text{ fm})^2 (1 \text{ fm})} \simeq 11 \text{ GeV/fm}^3 \tag{1.1}$$

Normal nuclear matter is not easily compressible [14, 15] and we can compare the energy density of this normal nuclear matter to the energy density given in Eq. (1.1). The volume of a spherical nucleus of radius R is given by $V = (4\pi R^3)/3$. So, the energy density ($\epsilon = m/V$) of a proton, is $\epsilon \sim 0.38$ GeV/fm^3, while $\epsilon \sim 0.17$ GeV/fm^3 for ^{208}Pb. Comparing the energy density of a proton and ^{208}Pb shows that the order of magnitude of the energy density of normal nuclear matter doesn't change irrespective of how much normal nuclear matter is present. The energy density given by Eq. (1.1) for an ultrarelativistic heavy-ion collision system is two orders of magnitude larger than the energy density of ordinary nuclear matter, which is strong evidence that there must be something drastically different happening in these heavy-ion collisions.

Furthermore, we can also compare the temperature of a system of ordinary nuclear matter to a QGP system at the energy given in Eq. (1.1). The Equation of State for a massless ideal gas is given by Eq. (1.2) [16],

$$P = \frac{1}{3}\epsilon, \quad \epsilon = g\frac{\pi^2}{30}T^4 \tag{1.2}$$

where P gives the pressure of the system, ϵ is the energy density, T gives the temperature of the system and g is the effective number of degrees of freedom. If we assume that the fundamental degrees of freedom of the nuclear matter described by Eq. (1.1) are pions, then $g \sim 3$ [16, 17] for an ideal massless pion gas and the temperature of the medium is, $T \sim 500$ MeV. If we assume a weakly coupled ideal gas of QCD, the degrees of freedom for two flavour QCD are $g = 37$ [17] and the temperature of the medium with energy density given in Eq. (1.1) is, $T \sim 300$ MeV. In order to get a sense of what these temperature values mean, we compare them to lattice QCD calculations.

The energy density and pressure as a function of temperature from lattice QCD calculations are shown in Fig. (1.2) [16, 19]. These lattice calculations show that there is a rapid change of the energy density at $T \sim 190$ MeV. However, more recent lattice QCD calculations show

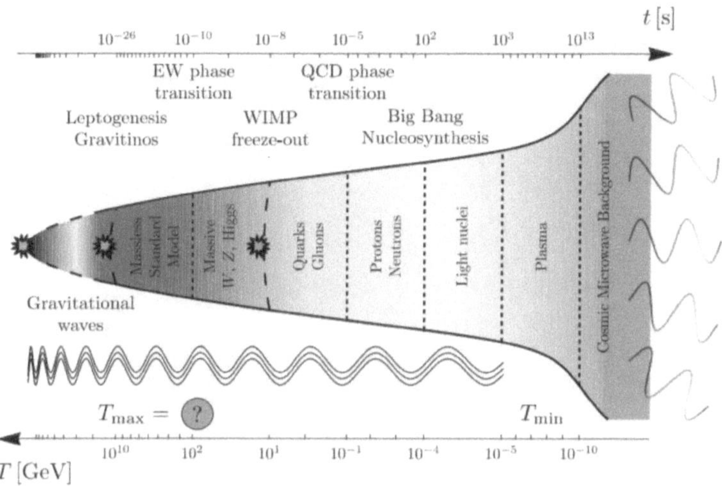

Figure 1.1: An approximate timeline of the universe obtained in Ref. [18].

that the QCD crossover takes place when $T \sim 155$ MeV [20]. The rapid change of the energy density and the fact that the temperatures corresponding to the energy density in Eq. (1.1) are much greater than $T \sim 155$ MeV, indicate that something interesting must be happening at $T \sim 155$ MeV. In the case of an ideal massless pion gas system, $\epsilon/T^4 = 3\pi^2/30 = 0.987$, while for a two flavour QGP system, $\epsilon/T^4 = 37\pi^2/30 = 12.2$. We see that ϵ/T^4 for two flavour QCD lies in the region beyond where this interesting event happens in Fig. (1.2). Therefore these lattice calculations [19] support the idea that something significant must be happening at $T \sim 155$ MeV. Therefore, we can conclude that a quark-gluon plasma (QGP) is indeed being created in ultrarelativistic collisions. A detailed discussion on other observables in heavy-ion collisions that support the formation of QGP can be found in Ref. [21]

Lattice calculations show that thermodynamic quantities such as the energy density shown in Fig. (1.2) deviate from the Stefan-Boltzmann limit of QCD (ideal massless gas or zero coupling limit) by $\sim 10\%$ [16]. This deviation from the Stefan-Boltzmann limit can be thought of being a result of the medium at $T \sim 190$ MeV being strongly coupled. Thus we can't use weak coupling techniques in this regime [16, 22]. This **book** is about trying to see if there exists signatures of this strong coupling physics imprinted on heavy quarks. By assuming that the quark-gluon plasma is strongly coupled, we use AdS/CFT [23–26] techniques to study how heavy flavoured particles lose their energy as they traverse through the QGP medium.

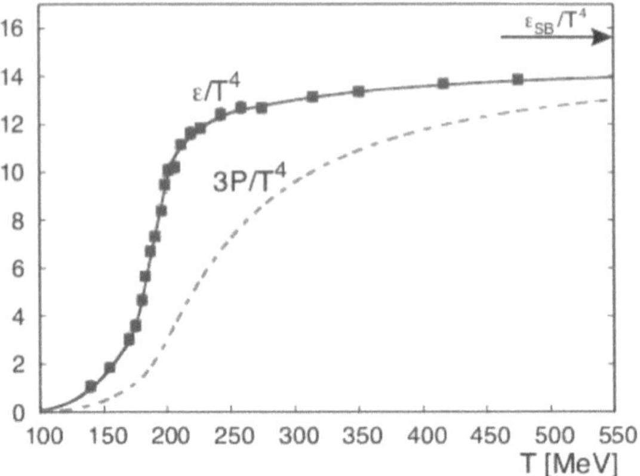

Figure 1.2: Results of the energy density ϵ/T^4 and pressure $3P/T^4$ as a function of temperature from lattice calculations. The arrow shows the Stefan Boltzmann limit of the energy density, obtained in Ref. [16].

1.2 Heavy-ion collisions and the QCD phase diagram

Another motivation for studying heavy-ion collisions theoretically and experimentally is to improve our understanding of the QCD phase diagram. We hope to find that the non-Abelian nature of QCD manifests itself in an interesting way in the QCD phase diagram, that can be experimentally measured and theoretically calculated. In Fig. (1.3) we show a sketch of our current conception of the QCD phase diagram as a function of temperature and the baryon chemical potential, from numerous theoretical investigations [27–31] and experimental measurements [32–34].

One of the major features of the QCD phase diagram is the quark-gluon plasma and its transition to hadronic matter. This region of the phase diagram is being explored by LHC and RHIC experiments at various collision energies ($\sqrt{s_{NN}}$). The numbers in the QGP region of Fig. (1.3) correspond to these various energies and the lines from the energies show the estimated trajectories of the droplets of QGP formed at the respective energies in heavy-ion collisions as they expand and cool [9]. The region of the phase diagram where $\mu_B = 0$ MeV and the low μ_B region is well understood from quantitative and controlled lattice calculations [36–40] as shown by the yellow region around $T \sim 155$ MeV [41, 42]. These lattice calculations at $\mu_B = 0$ MeV give a good approximation of the mid-rapidity LHC and RHIC data [9] because at high collision energies, the nuclei collide at speeds $\sim c$, and the hot QCD medium they produce has

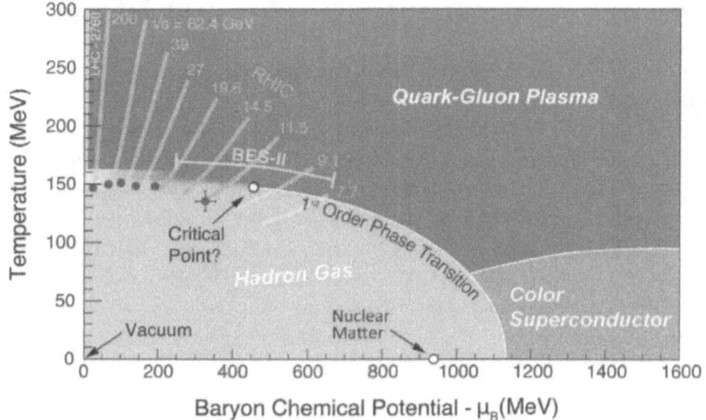

Figure 1.3: A sketch of the QCD phase diagram given as a function of temperature and the baryon chemical potential obtained in Ref. [35].

no net baryon number (i.e. $\mu_B = 0$ MeV).

The lattice calculations show that in the low μ_B region, the transition from QGP to hadronic matter is a continuous crossover [9, 31, 43, 44]. As μ_B is increased, lattice calculations become extremely difficult [45, 46]. There are some models that have been proposed which feature a critical point in which the transition from QGP to hadronic matter changes from a continuous crossover to a first order phase transition at some $\mu_B > 0$ MeV [47]. However, there are no first-principles theoretical calculations which illustrate the existence of a critical point in the QCD phase diagram or where it's located in the (μ_B, T) plane [47–49].

The work of this book seeks to contribute towards the use of relativistic heavy-ion collisions to improve our understanding of the QCD phase diagram. In particular, our studies are focused on the region of low baryon chemical potential (i.e $\mu_B \sim 0$ MeV) and temperatures in the order of a few hundreds MeV. By looking at the sketch of the QCD phase diagram in Fig. (1.3), this region is shown by the line closest to the temperature axis representing LHC collision energies. Lattice calculations can describe thermodynamic quantities [28, 38, 46] in this region, however, they cannot be used to describe real time processes such as the motion of a heavy quark through the QGP medium. There are very good reasons to believe that QCD is strongly coupled in this region and we'll discuss some of these reasons in the next two sections. If the theory is strongly coupled in this region, weak-coupling QCD techniques cannot be used either, hence we've used AdS/CFT techniques in this book.

1.3 Stages of relativistic heavy-ion collisions

There are several stages of a relativistic heavy-ion collision, starting from the initial colliding nuclei to the point where experiments can detect the various hadrons that are formed. In Fig. (1.4), we show a schematic view of the stages of a relativistic heavy-ion collision. Each of these stages is very complex and is subject to ongoing research with a lot of open questions [9] such as: "What are the qualitative differences, between the description of hydrodynamisation in a heavy ion collision modelled by assuming a weakly coupled initial stage (perturbative calculations take $\alpha_s = 0.3$) versus a strongly coupled holographic calculation (these treat the corresponding 't Hooft coupling $\lambda \approx 11$)". We will return to this discussion in Chap. (4) when we look at the two different set of parameters used in this book to account for the systematic theoretical uncertainties associated with the mapping of parameters in $\mathcal{N} = 4$ SYM theory to QCD.

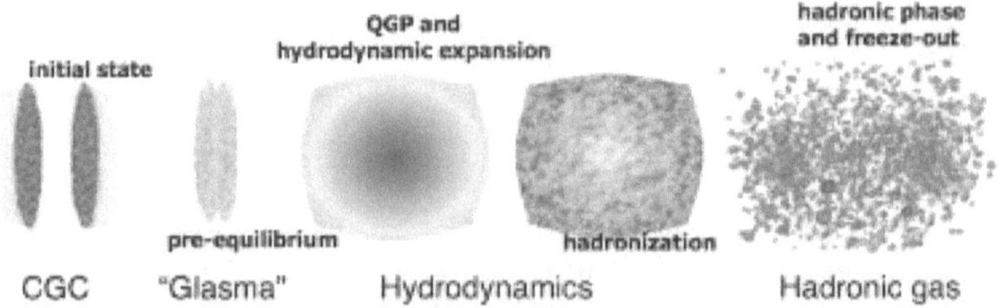

Figure 1.4: A schematic view of a relativistic heavy-ion collision, obtained in Ref. [50].

During a relativistic heavy-ion collision, we start with an incident nucleus with a diameter of approximately 14 fm (for a large nucleus such as ^{208}Pb) in its rest frame. It's then boosted in the laboratory frame such that its thickness is $\sim 14/\gamma$ fm, where γ is the Lorentz factor. The value of γ is approximately 100 at RHIC and 2500 for LHC energies [9, 51, 52]. The bunches of nuclei look like a disk due to Lorentz contraction and as the disks collide, QGP is formed and the incident quarks and antiquarks lose some of their energy but most of them don't experience a large angle change (i.e. most of them are 'soft' interactions, with little transverse momentum transfer). Some of the incident partons experience hard perturbative interactions and result in the production of particles that have high transverse momentum [9, 53], which will be of particular importance to the work of this book.

Particle accelerator detectors can be used to measure the charged particle spectrum from heavy-ion collisions [54]. At high transverse momentum, this spectrum is dominated by hadrons that originate from the fragmentation of hard scattered partons during the collision [55]. The suppression of high-p_T particles is an experimental observable indicating that these particles lose some of their energy as they interact with the QGP medium. The increase in collision energies at the LHC has resulted in the abundant production of high-p_T particles [56]. This increase in the production of high-p_T particles is expected to improve experimental constraints on energy loss models and also allow medium parameters to be determined more precisely [57]. It is important for us to create theoretical models that can be used to make predictions of this experimental observable. The work of this **book** will be focused on this task. Having such theoretical models also allows us to investigate the mechanism by which the hard partons lose their energy as they traverse the QGP [58, 59]. We now discuss the various stages of relativistic heavy-ion collisions.

Pre-equilibrium

The relativistic collision produces a dense and highly excited state of matter (fireball) made up of quarks and mostly gluons [50]. This system is not yet in equilibrium and the quarks and gluons frequently collide until a local equilibrium is established (thermalisation). The time taken to establish equilibrium is known as the thermalisation time and is approximately 1 fm/c [60]. It is not entirely understood how the system establishes equilibrium so fast [61].

Hydrodynamic expansion

Once the system has thermalised, the individual partons (quarks and gluons) remain in a deconfined state and the system has a thermal pressure, acting on the surrounding vacuum [50]. The transverse velocity of the QGP is small shortly after the collision (say at $\approx$ 1 fm/c), but it rapidly builds up to $\mathcal{O}(c)$ due to the pressure-driven hydrodynamic expansion [9]. The QGP has a very low viscosity to entropy density ratio, $\eta/s \approx 1/4\pi$ at approximately 1 fm/c after formation in the rest frame of the fluid [62, 63]. Leading order AdS/CFT results show a qualitative agreement of the timescale of thermalisation [64, 65] and the size of η/s [26] with values obtained from viscous relativistic hydrodynamics models and compared to RHIC and LHC data for $p_T \lesssim 1$ GeV particles [66].

As the two Lorentz contracted discs move away from each other after the collision, the QGP that has formed between them expands and cools, but as this happens, the formation of new QGP continues in the wake of each of the discs. The formation of new QGP is caused by the quarks and gluons produced at high rapidity as they are moving at very high speeds ($\sim c$) in one of the beam directions. Ultimately, when enough time has passed for particles produced at high rapidity to form QGP in their rest frame, a much longer period of time has passed in the lab frame, for example, for particles produced at rapidity $y = 6.5$, this time is around 330 fm/c [9]. The system goes through a collective expansion that can be described using relativistic hydrodynamics [67]. As it expands, the energy density drops and the system cools down. This expansion and cooling of the system is governed by energy-momentum conservation equations, involving an equation of state $p = p(\epsilon, n_B)$. As the temperature of the system goes down below a critical temperature $T_c \simeq 155$ MeV [20], the partons start to hadronise forming light particles such as pions.

Hadronisation

Hadronisation starts to occur approximately 10 fm/c after the collision [53]. During hadronisation, the entropy density decreases very fast over a very short period of time (entropy conservation holds) and over a narrow temperature range $T \approx 150 - 200$ MeV. The temperature dependence on thermodynamic variables such as the entropy density indicate that the effective degrees of freedom of the system change rapidly across the narrow temperature range [16]. Given that the total entropy cannot decrease, this means that the system expands rapidly while the temperature of the system remains approximately constant [50]. If the transition from the quark-gluon plasma to a hadron gas is first order, a mixed phase should exist, a state where the QGP and hadronic resonance gas can co-exist. However, if the phase transition is the second order one or there is no phase transition (crossover), there is no mixed phase. [68].

Hadronic phase and freeze out

After hadronisation, all the partonic matter in the QGP has been converted to hadronic matter, but the system remains in local thermal equilibrium [50]. The constituent hadrons continue to collide thus maintaining equilibrium while the system continues to expand and cool, we essentially have a hot hadron gas [53]. Eventually the inelastic collisions become too small to keep up with the expansion and the system reaches a chemical freeze-out [69] (the quantity of hadrons remains the same after chemical freeze-out). The system remains in local equilibrium and continues with its expansion (with possible elastic collisions) as it cools down, until the average distance between the constituent hadrons is larger than the strong interaction range [50]. At this point, the elastic collisions become less frequent to a point that local thermal equilibrium cannot be sustained. The hydrodynamic description breaks down at this point and the constituent hadrons decouple, this is called kinetic freeze-out [70], estimated to occur ~ 20 fm/c after the collision [53] and the hadrons are then picked up by the detectors.

1.4 Energy loss calculations

In Sec. (1.3), we discussed that high transverse momentum partons fragment into high transverse momentum particles (i.e. hadrons). These high-p_T particles are the most direct probe of the relevant degrees of freedom in a quark-gluon plasma [25, 71]. The idea is that these particles lose energy as they propagate through the QGP medium [72]. We can learn about the properties of the medium by making assumptions about how these high-p_T particles interact with the medium and making predictions of observables that can be compared to experimental data. One such observable is the nuclear modification factor, $R_{AA}(p_T)$ given by [73]

$$R_{AA}(p_T) \;=\; \frac{dN^{AA}/dp_T}{\langle N_{coll}\rangle dN^{pp}/dp_T} = \frac{dN^{AA}/dp_T}{T_{AA}d\sigma^{pp}/dp_T} \tag{1.3}$$

$$T_{AA} \;=\; \langle N_{coll}\rangle/\sigma^{pp}_{inel} \tag{1.4}$$

and is used to quantify this energy loss, where $\langle N_{coll}\rangle$ is the total number of binary nucleon-nucleon collisions, dN^{AA}/dp_T and dN^{pp}/dp_T are the measured particle spectra in AA and pp collisions respectively. The nuclear modification factor tells us how the nucleus-nucleus environment modifies the high-p_T particle production compared to a proton-proton environment [73]. In this book we'll study the energy loss of bottom quarks and consequently B-mesons as they propagate through the QGP medium.

Studies of the energy loss of quarkonia, high transverse momentum light or heavy hadrons from experimental measurements allow us to measure the physics of QGP. Generally, a typical energy loss calculation gives us the probability, $P(\Delta p_T | p_T, L, T, M_Q, R)$, of a parton of mass (or effective mass) M_Q losing some of its initial momentum as it propagates through a medium, i.e. a plasma of temperature T, where L gives the path-length the parent parton travels and R gives the representation (i.e. the parent parton; a quark or a gluon) [71]. This probability can be used to provide an approximation of the nuclear modification factor and this is discussed extensively in Ref. [71, 74].

One way of conceptualising how high-p_T particles interact with the medium is via the weak coupling picture. As the heavy quarks propagate through the QGP medium, they scatter off the various constituents of the medium [75], leading to radiative [76–78] and collisional energy loss [75]. Radiative processes include medium-induced gluon emission [77] and these processes become relevant for high-p_T heavy quarks [79, 80]. Collisional energy loss describes the redistribution of energy between the incoming and outgoing particles and is mostly responsible for heavy quark thermalisation at low-p_T [81].

Another view is that of strong coupling and we're going to take this view for the rest of this book. Note that the relevant scale for the heavy quark energy loss problem is the typical momentum transfer during interactions and not the quark's transverse momentum [82]. The momentum transfer describes the energy exchanged between the heavy quark and the medium partons as it propagates through the QGP medium [83]. The momentum transfer also informs the diffusion of the heavy quark propagating through the QGP medium [84]. In processes in-volving a small momentum transfer, non-perturbative corrections become important, but they are impossible to calculate using weak coupling techniques [16]. In this regime, QCD matter is strongly coupled [25] and we resort to AdS/CFT [23–26] techniques to perform energy loss calculations. Another motivation for the use of these AdS/CFT techniques in energy loss calculations is that AdS/CFT gives a very good description of the shear viscosity to entropy density ratio, $\eta/s = 1/4\pi$ for QGP systems [85–87]. Extensive discussions on heavy quark energy loss models in a strongly coupled plasma can be found in Ref. [88–93].

Some reasons why heavy flavour observables [94, 95] are particularly interesting is because the quarks are massive ($m_Q \gg \Lambda_{QCD}$), the quarks are produced very early in the collision and their initial production can be described by next-to-leading order perturbative Quantum Chromodynamics (pQCD) [96–98]. They act as identifiable test particles (ideal probes), since they navigate the whole evolution of the QGP medium as they participate in and are affected by its dynamics, but remain conserved [99, 100]. Heavy flavour acts as a tomographic probe [56, 101, 102] of quark-gluon plasma and provides a well-defined physical scale in which the temperature of the QGP medium can be quantified through the pattern of quark diffusion.

It is important to compare our theoretical predictions to a range of experimental data, for example, by looking at the suppression of heavy flavour at different energies (i.e. RHIC and the LHC) through the nuclear modification factor [103]. This **book** will present energy loss predictions made at two different LHC energies, $\sqrt{s_{NN}} = 2.76$ TeV and $\sqrt{s_{NN}} = 5.5$ TeV. Some early results of the energy loss in the higher order strong coupling regime (AdS/CFT calculations) have shown favourable results of the nuclear modification factor, $R_{AA}(p_T)$ that is observed for electrons from heavy flavour decay at RHIC [71, 104] but generally over-suppressed $R_{AA}(p_T)$ for D mesons at the LHC by a factor of approximately five [71, 105].

There are energy loss models for both light and heavy flavoured particles that are informed by calculations that use leading order pQCD and have shown a broad qualitative agreement with both RHIC and LHC data [71, 103, 106]. On the other hand, energy loss models using AdS/CFT for early light flavour [107] and leading order heavy flavour [108] suggest a massive over-suppression of high-p_T particles compared to data. There is a lot of research focused on improving the AdS/CFT treatment of high transverse momentum observables. Examples include using AdS/CFT to model the medium only [89, 109, 110] or using AdS/CFT to model only a part of the energy loss [111]. More recent work such as Ref. [66] showed that by improving the strong coupling jet prescription and re-normalising the in-medium energy loss, a reasonable value of the 't Hooft's coupling $\lambda = 5.5$ produce a jet nuclear modification factor that is quantitatively consistent with preliminary CMS data. In addition, Ref. [88] showed an improved energy loss model including both drag and diffusion building on the work of Ref. [112]. In Fig. (1.5) we show CMS results for the suppression of B-mesons at $\sqrt{s_{NN}} = 5.02$ TeV including AdS/CFT predictions with a diffusion coefficient that is dependent on momentum and a diffusion coefficient that is constant, discussed in Ref. [88, 113]. This **book** will build on this approach.

In the energy loss model we've employed, heavy quarks are produced individually in the geom-

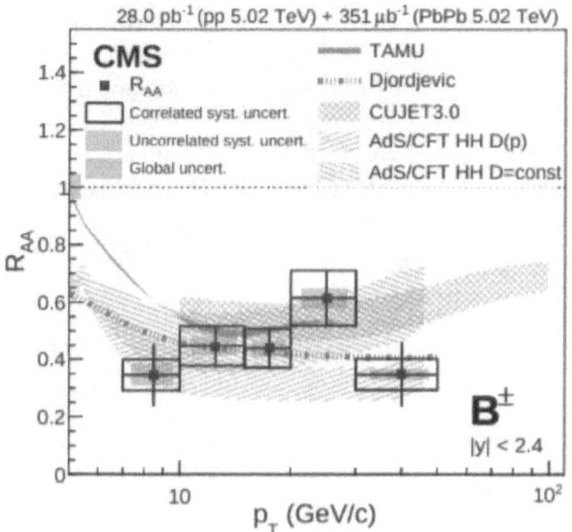

Figure 1.5: B-meson nuclear modification factor as a function of p_T at $\sqrt{s} = 5.02$ TeV [114].

etry, at thermalisation time, the QGP medium forms and they propagate through and interact with this medium, then they fragment. The QGP medium that we're using is a relativistic hydrodynamic background generated by VISHNU [115, 116]. The heavy quarks propagate through a 'different looking' medium depending on the angle in which they are produced, for example, the medium looks different for the various centrality classes. This difference in the medium results in the suppression of these heavy quarks having an azimuthal dependence. In principle, one can Fourier decompose any periodic function of an angular variable into its Fourier modes [117]. We can express the Fourier expansion of quantities such as the momentum integrated invariant distribution of particles in the form given in Eq. (1.5) [68, 118],

$$\frac{dN}{d\phi} = \frac{N}{2\pi} \left(1 + 2 \sum_{n=1}^{\infty} v_n \cos[n(\phi - \psi)] \right) \tag{1.5}$$

where ϕ is the azimuthal angle of the detected particles and ψ gives the angular orientation of the reaction plane. Using this approach, we Fourier expand our nuclear modification factor, $R_{AA}(p_T, \phi)$ result obtained in the energy loss model as shown in Eq. (1.6),

$$R_{AA}(p_T, \phi) = R_{AA}(p_T) \left[1 + 2v_2(p_T) \cos(2\phi) \right] \tag{1.6}$$

and we extract the coefficient $v_2(p_T)$.

Notice that the Fourier coefficient, $v_2(p_T)$ has a momentum dependence. There are multiple orders of magnitude more low transverse momentum particles produced in heavy-ion collisions compared to the high-p_T particles. The high transverse momentum particles are a result of incident partons that experience hard perturbative interactions, while the low momentum particles are coming from the medium (i.e. soft processes). Consequently, the relevant physics when we have fewer particles is going to be different from the relevant physics when we have orders of magnitude more particles. In Fig. (1.6) we show the $v_2(p_T)$ for B and D mesons [88] computed using this approach at $\sqrt{s} = 2.76$ TeV and compared to ALICE results [119].

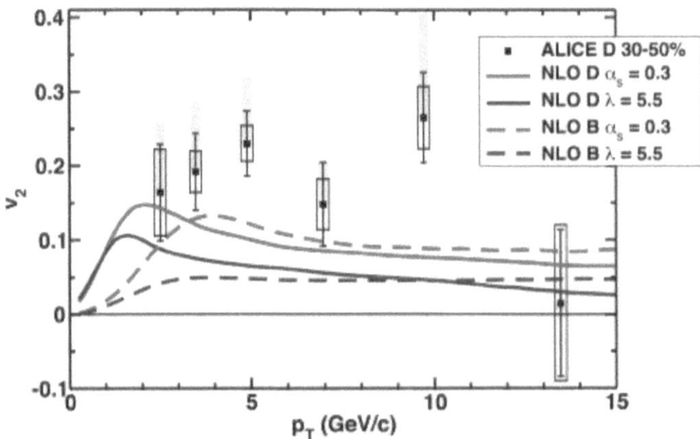

Figure 1.6: B and D meson $v_2(p_T)$ extracted from the Fourier expansion of the nuclear modification factor obtained in Ref. [88].

The main result of this **book** will be the nuclearmodification factor, $R_{AA}(p_T)$, as well as the corresponding $v_2(p_T)$ for B mesons at both $\overline{s_{NN}} = 2.76$ TeV for central collisions and $\sqrt{s_{NN}} = 5.5$ TeV for various centrality classes (i.e. central, semi-central and peripheral collisions). The optical limit of the Glauber model, which we use in our initial conditions (i.e. to model the geometry of the colliding nuclei) is discussed in Chap. (2) and a comparison of some geometric quantities is made to the Glauber Monte Carlo approach. In Chap. (3), we discuss the statistical uncertainties associated with the use of a Monte Carlo random number generator [120] to produce the heavy quarks at specific points in the hydrodynamic medium. The Langevin energy loss model, which describes the motion of a heavy quark strongly coupled to a QGP is discussed in Chap. (4). In Chap. (5) we go through the methods used to compute the nuclear modification factor as well as the $v_2(p_T)$ and the corresponding statistical uncertainties, then we give the main results of the book in Chap. (6). Throughout the book, the following units will be used: p_T (GeV/c) and t (fm/c).

Chapter 2

Collision Geometry with the Optical Glauber Model

In this chapter we will discuss the modelling of the initial geometry of a heavy-ion collision using the optical Glauber model [121]. We use the optical limit of the Glauber model to estimate the collision probability density (i.e. the density of binary collisions) in the transverse plane. Then we use the collision probability density as a probability density function in a Monte Carlo random number generator [120]. This random number generation process allows us to produce heavy quarks at specific points in the hydrodynamic medium and will be discussed in detail in Chap. (3).

2.1   Input parameters

The Glauber model is used to compute the geometric quantities that are used for initial conditions of a heavy-ion collision, that is, it models the geometry of the nuclei before an event. As inputs in the computation of geometric quantities, the Glauber model takes in some experimental parameters such as the nuclear charge density of the colliding nuclei as well the inelastic nucleon-nucleon cross section.

2.1.1   Nuclear charge density

The nucleon charge density (inside the nucleus) is given by the Fermi distribution [121, 122],

$$\rho(r) \;=\; \frac{\rho_0(1 + w(r/R)^2)}{1 + exp\left[\frac{r-R}{a}\right]} \tag{2.1}$$

where ρ_0 is the density of the nucleon at the centre of the nucleus, $r = \sqrt{x^2 + y^2 + z^2}$ is the distance from the centre of the nucleus, R is the radius of the nucleus, a is the 'skin depth' (it measures how quickly $\rho(r)$ falls off at the surface of the nucleus) and w describes deviations of the nucleus from a spherical shape. The following table shows values of these parameters for different nuclei and in this **book** we will study $Pb + Pb$ collisions.

Table 2.1: Nuclear density parameters [123]

Nuclei	R (fm)	a (fm)
^{63}Cu	4.20 ± 0.02	0.596 ± 0.008
^{197}Au	6.38 ± 0.06	0.535 ± 0.027
^{208}Pb	6.624 ± 0.035	0.549 ± 0.008

Assuming $w = 0$, this density assumes the Woods-Saxon distribution [124],

$$\rho(r) \quad = \quad \frac{\rho_0}{1 + exp\left[\frac{r-R}{a}\right]} \tag{2.2}$$

Normalisation

For different nuclei, the nucleon density at the centre of the nucleus (ρ_0) is fixed by the following normalisation condition [125, 126]:

$$\int \rho dV \quad = \quad \int_0^\infty 4\pi r^2 \rho(r) dr = 1 \tag{2.3}$$

For a hard-sphere configuration we have:

$$\rho(r) = \begin{cases} \rho_0, & \text{if } r < R \\ 0, & \text{if } r \geq R \end{cases} \tag{2.4}$$

thus giving

$$\rho_0 \quad = \quad \frac{3}{4\pi R^3} \tag{2.5}$$

A discussion on the density distribution of deformed nuclei can be found on Ref. [124]. The ratio of the nucleon charge density (as given by the Woods-Saxon distribution) as a function of the distance from the centre of the nucleon to the nucleon density at the centre of the nucleus is shown in Fig. (2.1) for different nuclei. The Woods-Saxon distribution parameters are given in Table (2.1) for the nuclei under consideration and a broader discussion on nuclear charge density distribution parameters can be found in Ref. [127, 128].

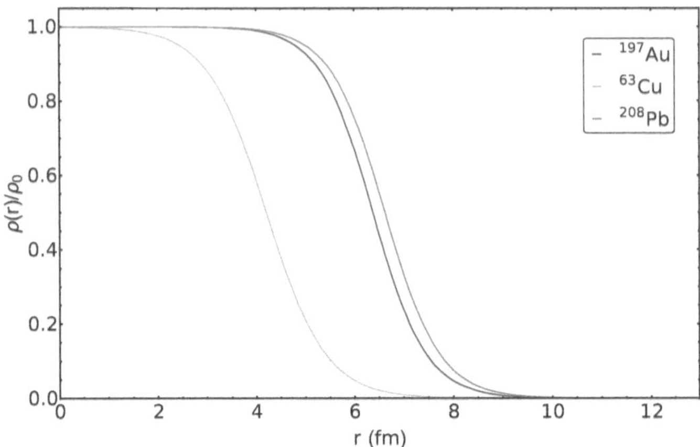

Figure 2.1: Density distribution for different nuclei used in heavy-ion collisions with parameters given in Table (2.1)

2.1.2 Inelastic nucleon-nucleon cross section

Heavy-ion collisions involve multi-particle nucleon-nucleon processes and one of the assumptions in the Glauber model is that there are no elastic collisions between the nucleons and that the average number of charged particles that are produced during each of the collisions is constant [124]. During these collisions, there's a small change in momentum and the collision cross section is taken to be the same as that of a single proton-proton collision. Glauber-model calculations do not consider elastic processes since they lead to very little energy loss.

Perturbative quantum chromodynamics (pQCD) calculations have good applicability for transverse momentum, $p_T \gtrsim 1$ GeV/c [129]. These pQCD calculations can't be applied when determining the inelastic nucleon-nucleon cross section (σ_{inel}^{NN}) since the cross section also accounts for processes involving a low momentum transfer [121]. As a result, the model takes in as input, the experimental measured cross section data which gives the beam-energy dependence of Glauber calculations [130]. The following table gives the inelastic nucleon-nucleon cross section for collision energies appropriate for RHIC, LHC and the FCC. The last column includes appropriate units, as used in our computations.

Table 2.2: Inelastic nucleon-nucleon cross section (σ_{inel}^{NN}) for collision-energies ($\sqrt{s_{NN}}$) appropriate for RHIC, LHC and the FCC [123]

$\sqrt{s_{NN}}$ (TeV)	σ_{inel}^{NN} (mb)	σ_{inel}^{NN} (fm^2)
0.2	41.6 ± 0.6	4.16
0.9	52.2 ± 1.0	5.22
2.76	61.8 ± 0.9	6.18
5.02	67.6 ± 0.6	6.76
5.44	68.4 ± 0.5	6.84
5.5	68.5 ± 0.5	6.85
7	70.9 ± 0.4	7.09
8	72.3 ± 0.5	7.23
8.16	72.5 ± 0.5	7.25
8.8	73.3 ± 0.6	7.33
10.6	75.3 ± 0.7	7.53
13	77.6 ± 1.0	7.76
14	78.4 ± 1.1	7.84
17	80.6 ± 1.5	8.06
27	86.0 ± 2.4	8.60
39	90.5 ± 3.3	9.05
63	96.5 ± 4.6	9.65
100	102.6 ± 6.0	10.26

2.2 Geometric quantities in the optical limit

The optical limit approximation of the Glauber model assumes the following [124]:

- at high energies, the nucleons carry a sufficiently large momentum such that they do not deflect as the nuclei pass through each other

- nucleons move independently in the nucleus

- the size of the nucleus is large relative to the extent of the nucleon-nucleon force

As a result of these approximations, for calculations, the nucleus is assumed to comprise of a smooth/continuous nucleon density that is given by the Woods-Saxon distribution in the radial direction and it's uniform over the polar and azimuthal angles [124]. The optical limit theory doesn't identify the individual nucleons at the respective spatial coordinates as in the Glauber Monte Carlo approach, discussed in Sec. (2.3). Consequently, analytic expressions for various geometric quantities can be easily developed by assuming that the constituent nucleons follow independent linear trajectories.

2.2.1 Formalism

We now discuss the set-up of the geometry of the colliding nuclei. Consider two Lorentz contracted discs of heavy-ion beams (target A and projectile B) colliding at relativistic speeds at impact parameter b as shown by the sketch in Fig. (2.2). The reaction plane in the xz-axes is defined by the beam direction as well as the impact parameter and the transverse plane is in the xy-axes. Instead of considering a vector impact parameter ($\boldsymbol{b}$), we replace it by a scalar distance (b) by assuming that the colliding nuclei are not polarised [121].

In setting up equations to describe the overlap region of the two colliding disks, we've used two different approaches to shift the each of the disks in the axis of the impact parameter and this is related to defining an origin. The first approach is shifting one of the disks by $(-b/2, 0, 0)$ and the other by $(b/2, 0, 0)$. The second approach is shifting one disk by $(b, 0, 0)$ and not shifting the other. In this **book**, we've primarily used the first approach, however, we've also used the second approach to check for consistency in our computational method since these two approaches should yield the same result.

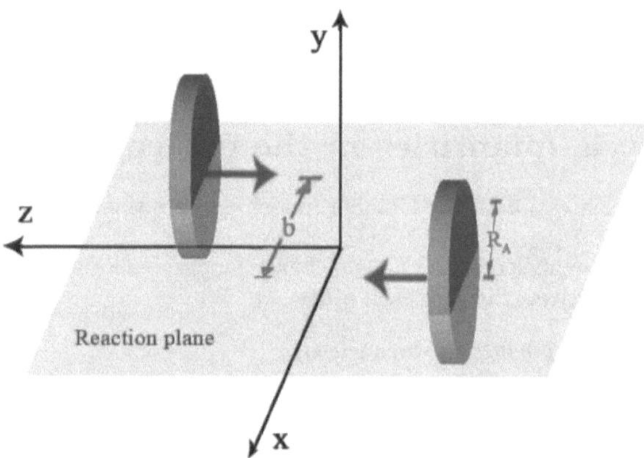

Figure 2.2: A collision of two heavy ions each with nuclei radius R_A and R_B at impact parameter b [131].

The points of interest are the two flux tubes (one on each nuclei) that overlap during the collision. One is located at a point $(x - b/2, 0, 0)$ with respect to the centre of nucleus A and the other is at $(x + b/2, 0, 0)$ from the centre of nucleus B. One finds that the probability per unit transverse area of a particular nucleon being located in the flux tube of either nucleus A or B (called the thickness function) is given by [123],

$$T_{A/B}(x, y) \;=\; \int_{-\infty}^{\infty} \rho(x, y, z_{A/B}) dz_{A/B} \qquad (2.6)$$

where $\rho(x, y, z_{A/B})$ is the probability per unit volume, of finding a nucleon at a point $(x, y, z_{A/B})$ in the nucleus of projectile (A) or target (B). This probability per unit volume is normalised to unity as discussed in Sec. (2.1). The combined probability per unit transverse area of finding nucleons located in the respective overlapping flux tubes of nucleus A and B is given by what is defined as the nuclear overlap function, $T_{AB}(b)$ given by Eq. (2.7) [123].

$$T_{AB}(b) \;=\; \int T_A\left(x - \frac{b}{2}, y\right) T_B\left(x + \frac{b}{2}, y\right) dx dy \qquad (2.7)$$

The nuclear overlap function is a geometric factor and has units of inverse area (i.e. fm^{-2}), Fig. (2.3) shows the overlapping region of the colliding nuclei. Multiplying the nuclear overlap function with the inelastic nucleon-nucleon cross section, $T_{AB}(b)\sigma_{inel}^{NN}$ gives the probability of an interaction occurring.

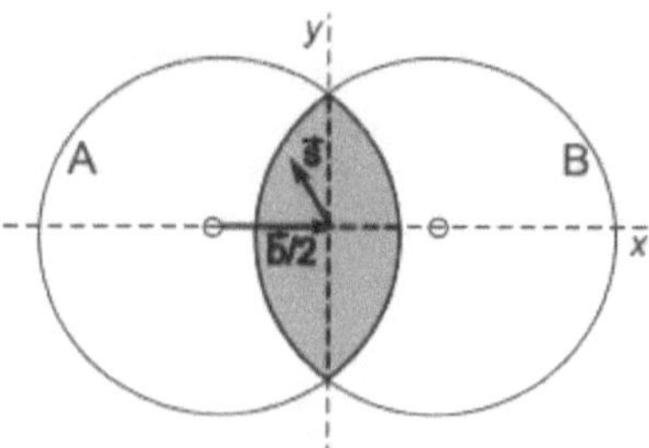

Figure 2.3: Nuclei A and B overlapping nuclei in the transverse plane. The vector $\vec{s}$ locates the position of the flux tube on nucleus A and B [132].

2.2.2 Geometric quantities of interest

We can now compute various geometric quantities of interest. We start by defining the binary nucleon-nucleon collision probability density, $n_{BC}(x, y)$ in the transverse plane given by [121]

$$n_{BC}(x, y; b) \quad = \quad AB\sigma_{inel}^{NN}T_A\left(x - \frac{b}{2}, y\right)T_B\left(x + \frac{b}{2}, y\right) \tag{2.8}$$

The binary nucleon-nucleon collision probability density allows us to find production points of the heavy quarks in the hydrodynamic medium for collisions at various impact parameters. We use this collision probability density as a probability density function for a Monte Carlo random number generator [120]. This random number generation process and the associated statistical uncertainties will be discussed extensively in the following chapter.

Using this collisions density, one can also compute the total number of binary nucleon-nucleon collisions as a function of the impact parameter by integrating out the transverse coordinates as shown in Eq. (2.9) [121].

$$N_{coll}(b) \quad = \quad \int n_{BC}(x, y; b)dxdy = ABT_{AB}(b)\sigma_{inel}^{NN} \tag{2.9}$$

Similarly, we define the number of participants (also known as wounded nucleons) in Eq. (2.10) [123],

$$N_{part}(b) \quad = \quad A\int T_A^-(1 - [1 - \sigma_{inel}^{NN}T_B^+]^B)dxdy + B\int T_B^+(1 - [1 - \sigma_{inel}^{NN}T_A^-]^A)dxdy \tag{2.10}$$

$$T_i^\pm \quad = \quad T_i\left(x \pm \frac{b}{2}, y\right) \ , \ \text{i=A or B} \tag{2.11}$$

which is the number of nucleons in each of the colliding nuclei that interacted at least once during the collision.

In Fig. (2.4) we show the number of binary nucleon-nucleon collisions and the number of participants as a function of impact parameter (b) in the optical limit of the Glauber model for $Pb + Pb$ at $\sqrt{s_{NN}} = 5.5$ TeV. The value of N_{part} is dependent on σ_{inel}^{NN} as shown in Eq. (2.10) and has an upper bound of $208 + 208 = 416$ at $b = 0$ fm for ^{208}Pb.

The geometric cross section distribution with respect to the impact parameter $d\sigma/db$ is given by Eq. (2.12) [126].

$$\frac{d\sigma}{db}(b) \quad = \quad 2\pi b(1 - [1 - T_{AB}(b)\sigma_{inel}^{NN}]^{AB}) \tag{2.12}$$

In Fig. (2.5) we show the geometrical cross section as a function of impact parameter (b) for $Pb + Pb$ at $\sqrt{s_{NN}} = 5.5$ TeV.

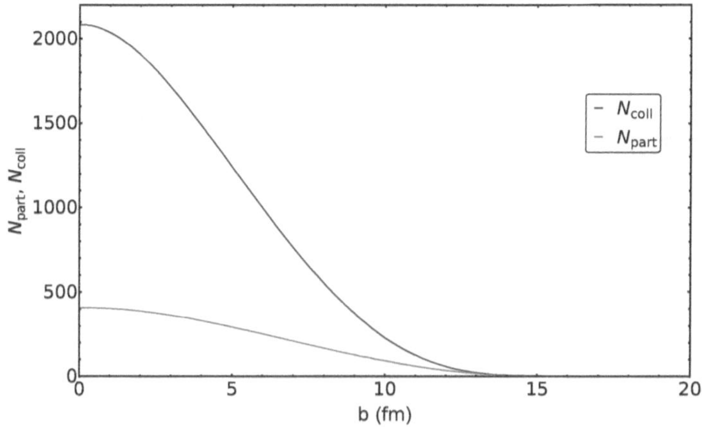

Figure 2.4: Number of collisions (N_{coll}) and the number of participants (N_{part}) with respect to the impact parameter (b) in the optical limit of the Glauber model for $Pb+Pb$ at $\sqrt{s_{NN}} = 5.5$ TeV

2.2.3 Centrality classes

The total geometric cross section (σ_{geo}) can be computed by taking the integral of the distribution $d\sigma/db$ given by Eq. (2.12) over all impact parameter as shown in Eq. (2.13).

$$\sigma_{geo} = \int_0^\infty \frac{d\sigma}{db} db \qquad (2.13)$$

This geometric cross section approximates the total inelastic cross section in ultra-relativistic heavy-ion collisions. In the framework of the Glauber Monte Carlo (GMC), one can think of this integral as the number of all GMC events where a minimum of one inelastic nucleon-nucleon collision occurs. Using Eq. (2.13), we integrate Fig. (2.5) over the impact parameter and obtain a total geometric cross section, $\sigma_{geo}^{PbPb} \approx 7917.6$ mb for $Pb+Pb$ at $\sqrt{s_{NN}} = 5.5$ TeV.

Centrality is a measure of how far apart the centres of two colliding nuclei are in the plane perpendicular to the collision axis. An illustration of how the area of the overlap region during the collision changes when moving from peripheral to central collisions in shown in Fig. (2.6).

Centrality classes are defined theoretically by taking slices of the impact parameter (b) distribution, shown in Fig. (2.5). These slices should be taken at b_{min} and b_{max} for each centrality class, where b is the impact parameter. Notice that $b = 0$ fm will be b_{min} for the $0 - X\%$

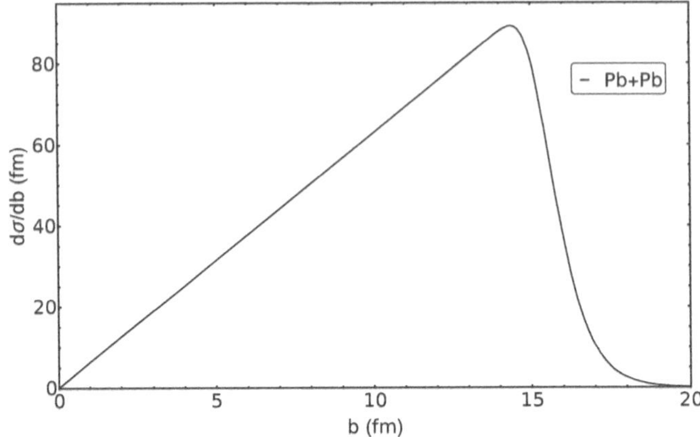

Figure 2.5: Total geometrical cross section with respect to the impact parameter (b) in the optical Glauber model for $Pb + Pb$ at $\sqrt{s_{NN}} = 5.5$ TeV

centrality class, so in order to find b_{max} for the $0 - 5\%$ centrality class, we need to solve

$$\int_0^{b_{0-5\%}} \frac{d\sigma}{db} db = \sigma_{geo} \times \frac{5}{100} = \frac{\sigma_{geo}}{20} \tag{2.14}$$

for the upper limit of the integral. Similarly, to find b_{max} for the $0 - 10\%$ centrality class, we need to solve for the upper limit of the integral in Eq. (2.15).

$$\int_0^{b_{0-10\%}} \frac{d\sigma}{db} db = \sigma_{geo} \times \frac{10}{100} = \frac{\sigma_{geo}}{10} \tag{2.15}$$

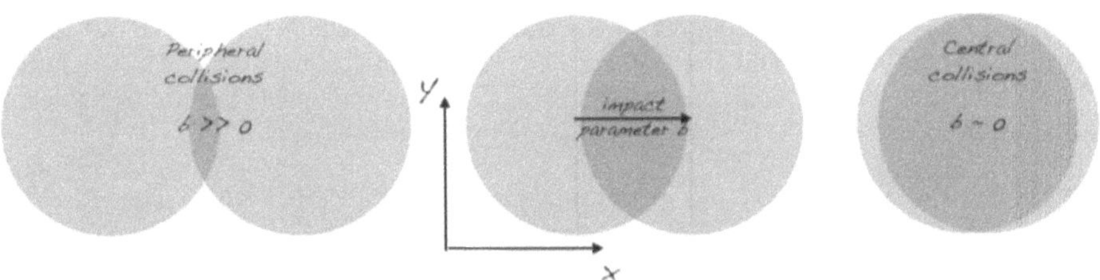

Figure 2.6: Centrality dependence on the impact parameter in the transverse plane during a heavy-ion collision, obtained in ref [133].

Then to find b_{max} for the $10-20\%$ centrality class, we need to solve for the upper limit of the integral in Eq. (2.16).

$$\int_{b_{0-10\%}}^{b_{10-20\%}} \frac{d\sigma}{db} db = \sigma_{geo} \times \frac{10}{100} = \frac{\sigma_{geo}}{10} \tag{2.16}$$

In our theoretical calculations, we use the average value of b (i.e. $\langle b \rangle$) for the respective centrality class which is given by;

$$\langle b \rangle_{0-10\%} = \frac{\int_0^{b_{0-10\%}} b \frac{d\sigma}{db} db}{\int_0^{b_{0-10\%}} \frac{d\sigma}{db} db} \tag{2.17}$$

$$\langle b \rangle_{10-20\%} = \frac{\int_{b_{0-10\%}}^{b_{10-20\%}} b \frac{d\sigma}{db} db}{\int_{b_{0-10\%}}^{b_{10-20\%}} \frac{d\sigma}{db} db} \tag{2.18}$$

Table (2.3) shows the centrality classes for $Pb+Pb$ at $\sqrt{s_{NN}} = 5.5$ TeV and they are comparable to results found in [123] using a slightly different approach. Note that there is a discrepancy in the value of $\langle b \rangle$ quoted in Table (2.3) for the 5-10% centrality class and the value used in the computations of this book. A value of $\langle b \rangle = 3.35$ fm was used for the computations (which corresponds to the 0-10% centrality class). We anticipate that this will have a very small impact on the $R_{AA}(p_T)$ and $v_2(p_T)$ results for this centrality class and the correction will be incorporated into Ref. [134].

Table 2.3: Centrality classes for $Pb + Pb$ at $\sqrt{s_{NN}} = 5.5$ TeV

Centrality	b_{min} (fm)	b_{max} (fm)	$\langle b \rangle$ (fm)
0-5%	0	3.55	2.37
5-10%	3.55	5.02	4.33
10-20%	5.02	7.01	6.12
20-30%	7.01	8.70	7.92
30-40%	8.70	10.04	9.38
40-50%	10.04	11.23	10.64
50-60%	11.23	12.30	11.77
60-70%	12.30	13.28	12.80
70-80%	13.28	14.20	13.75
80-90%	14.20	15.12	14.65
90-100%	15.12	21.69	15.95

2.3 Comparison with the Glauber Monte Carlo

In the Glauber Monte Carlo approach, the individual nucleons of each nucleus are randomly distributed event-by-event in a three-dimensional coordinate system where the distribution follows the appropriate nuclear density [124]. Then the impact parameter is randomly drawn from the $d\sigma/db = 2\pi b$ distribution. The nucleus-nucleus collisions are modelled as sequences of binary nucleon-nucleon collisions that are independent of each other and the inelastic nucleon-nucleon cross-section is taken to be independent of any previous collisions each nucleon has experienced. A nucleon-nucleon collision will only occur if the transverse distance (d) between nucleons from the different nuclei satisfies [121];

$$d \leq \sqrt{\frac{\sigma_{inel}^{NN}}{\pi}} \tag{2.19}$$

Geometrical quantities such as N_{coll} and N_{part} are calculated by simulating a high number of nucleon-nucleon collisions and averaging over them. In Fig. (2.7) we show a comparison of N_{coll} and N_{part} for $Pb + Pb$ at $\sqrt{s_{NN}} = 5.5$ TeV computed using the Glauber Monte Carlo and using the optical limit approach.

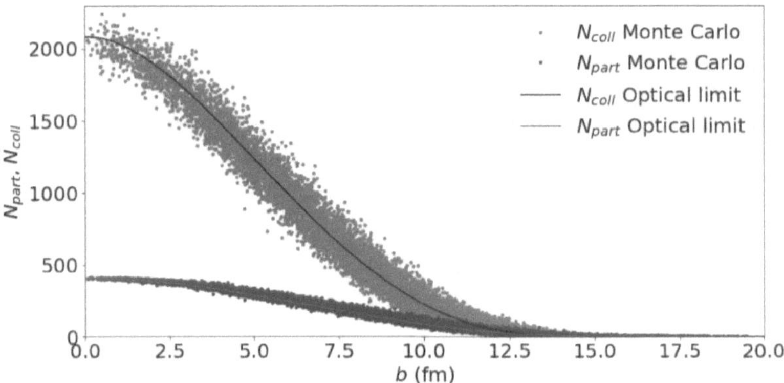

Figure 2.7: Comparison of N_{coll} and N_{part} for $Pb + Pb$ at $\sqrt{s_{NN}} = 5.5$ TeV computed using the Glauber Monte Carlo and using the optical limit approach

Chapter 3

Testing Production Geometry

The heavy-ion collision geometry can be described within the framework of the Glauber Model. The collision probability density at a specified impact parameter was given in Eq. (2.8). This collision probability density acts as a probability density function and allows us to produce heavy quarks at particular points in the hydrodynamic medium. We have used Monte Carlo methods [135–137] to generate a pair of pseudo-random numbers obeying the collision probability density given in Eq. (2.8). In particular, we use the *Ran* routine from Numerical Recipes [120]. This pair of random numbers gives us the production points of the heavy quarks in the hydrodynamic medium. Generating random numbers is discussed in Appx. (A) and more extensively in Ref. [120].

Generating the production points of the heavy quarks in the hydrodynamic medium is an extremely numerical process. It is important to have a quantitative way of testing to ensure that the heavy quarks are produced at appropriate points in the hydrodynamic medium. In this chapter, we will compute the statistical uncertainties associated with the use of Monte Carlo methods for heavy quark production points. We quantitatively compare the Monte Carlo distribution of pseudo-random numbers to the analytical result of Eq. (2.8). We will consider two cases; in the first case, we integrate out the y-coordinate in Eq. (2.8) to produce heavy quarks along the x-direction. One could also integrate out the x-coordinate and produce heavy quarks along the y-direction. In the second case, we produce heavy quarks in the transverse plane, shown in Fig. (2.6). In this case, the probability density function follows the two dimensional collision probability density described by Eq. (2.8).

The statistical uncertainties are propagated using Poisson statistics as follows;

$$\sigma \;=\; \sqrt{n\left(1 - \frac{n}{N}\right)} \tag{3.1}$$

where n is the total number of random numbers in each bin and N is the total number of random numbers produced.

3.1 Heavy quarks produced along a line

As a test case, we have integrated out the y-coordinate in Eq. (2.8) in order to produce preudo-random numbers obeying a one dimensional probability density function. This simpler case gives us a feel of the statistical uncertainties associated with generating pseudo-random numbers obeying a one dimensional probability density function. We use the y-integrated collision probability density to compute the numerical probability density function. We obtain the cumulative distribution function by numerically integrating this one dimensional probability density function. Following the procedure described in Appx. (A) and Ref. [120], we generate random numbers obeying this y-integrated collision probability density. In principle, this case corresponds to producing heavy quarks in a "two dimensional QGP system" consisting of the beam axis and a direction perpendicular to that axis.

We show the collision probability density along the x-direction for a $Pb + Pb$ collision at $\sqrt{s_{NN}} = 5.5$ TeV for the 0-5% centrality class in Fig. (3.1). This y-integrated collision probability density is normalised by its integral in the domain of interest and labelled 'Normalised PDF for x'. The histogram shows the Monte Carlo (MC) distribution of random numbers obeying this collision probability density (this distribution is for a hundred thousand random numbers). The y-integrated collision probability density shown in Fig. (3.1) is interpolated at points corresponding to the centre of bins of the Monte Carlo distribution. To get a quantitative sense of how well our Monte Carlo random numbers represent the probability density function, we take the ratio of the bin heights to the normalised PDF. This ratio is shown in Fig. (3.2) and shows that there's a good agreement for random numbers created close to the point $x = 0$ fm with less than 20% difference. The ratio diverges from one for points produced further away from $x = 0$ fm. The main reason for this is that we have high statistics near the point $x = 0$ fm and lower statistics as we move further from that point, this can be seen by looking at the histogram in Fig. (3.1). We also see a ratio of zero after $|x| \approx 7.5$ fm due to bins with zero random numbers in them.

In Fig. (3.3) we show the collision probability density along the x direction for $Pb + Pb$ at $\sqrt{s_{NN}} = 5.5$ TeV for the 0-5% centrality class as well as the distribution of random numbers obeying this collision probability density, but now, for one million random numbers. As can be seen from this figure, there's a clear improvement from Fig. (3.1) in that the histogram now matches the collision probability density better. The reason for this is that we've produced a factor of ten more random numbers in this case and thus have improved our statistics compared to the previous attempt where we only had a hundred thousand random numbers. Looking at the ratio in Fig. (3.4), we see a better agreement extending further out in x than in Fig. (3.2). Also note that we now have fewer bins with zero heavy quarks in them and the size of the statistical uncertainties has decreased. Therefore we conclude that in order to accurately generate heavy quarks at random points (in the quark gluon plasma) obeying the collision probability density as given by the Glauber Model, we need to generate a high number of heavy quarks so as to ensure that we have high statistics.

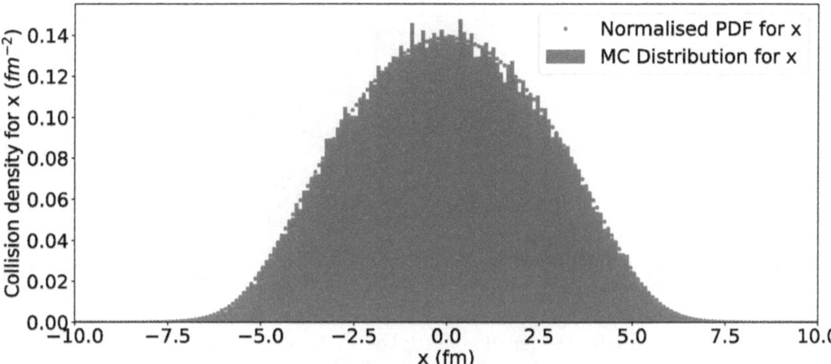

Figure 3.1: Collision probability density along the x direction for $Pb + Pb$, 0-5% centrality at $\sqrt{s_{NN}} = 5.5$ TeV as well as a histogram for 10^5 Monte Carlo generated random numbers obeying this distribution.

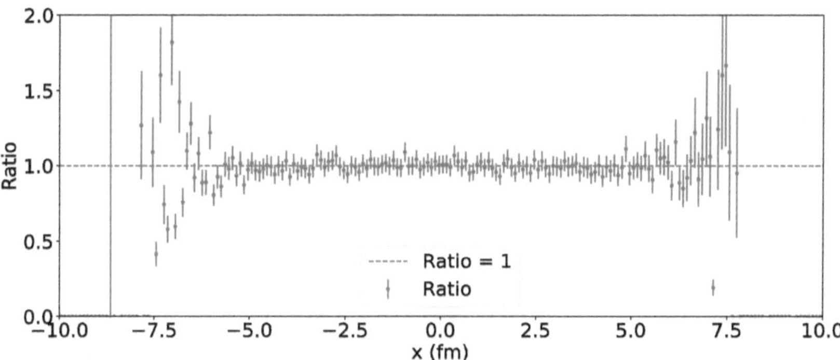

Figure 3.2: Ratio of the MC distribution to the collision probability density along x for $Pb+Pb$, 0-5% centrality at $\sqrt{s_{NN}} = 5.5$ TeV corresponding to the collision probability density plot shown in Fig. (3.1).

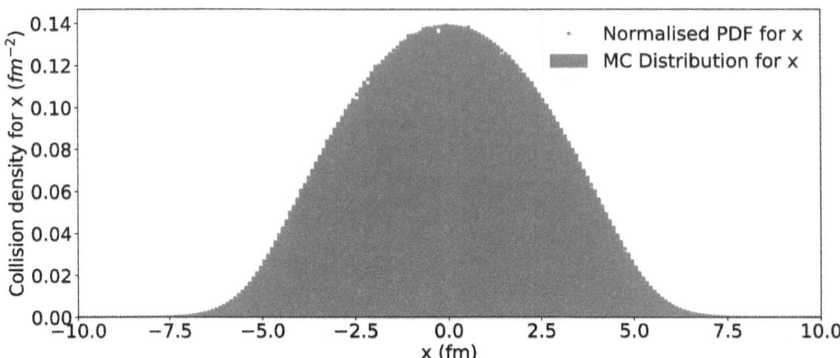

Figure 3.3: Collision probability density along the x direction for $Pb + Pb$, 0-5% centrality at $\sqrt{s_{NN}} = 5.5$ TeV as well as a histogram for 10^6 Monte Carlo generated random numbers obeying this distribution.

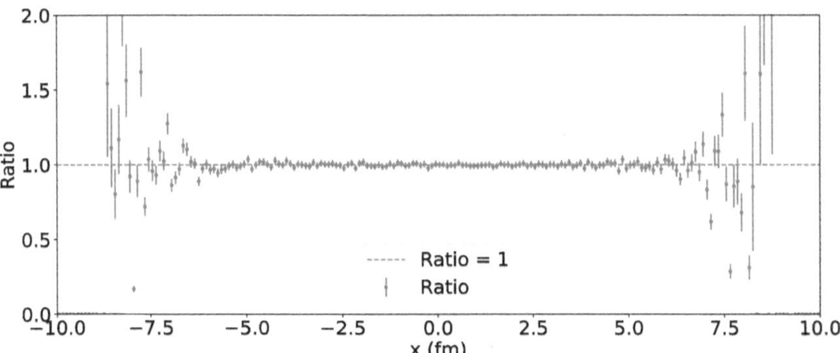

Figure 3.4: Ratio of the MC distribution to the collision probability density along x for $Pb+Pb$, 0-5% centrality at $\sqrt{s_{NN}} = 5.5$ TeV corresponding to the collision probability density plot shown in Fig. (3.3).

The collision probability density along x that we've looked at so far is only for the 0-5% centrality class, similarly we can do the same for other centrality classes. In Fig. (3.5) we show the collision probability density along x for $Pb + Pb$ at $\sqrt{s_{NN}} = 5.5$ TeV for various centrality classes. As expected, we see that the collision probability density becomes less spread and strongly peaked around $x = 0$ fm as we move from central to peripheral collisions. The reason for this change in the shape of the collision probability density is that the overlap area between the two nuclei during a collision is higher for central collisions and lower for peripheral collisions since there will be fewer participants and more spectators as illustrated by the sketch in Fig. (3.12).

The collision probability density along x has more variability for central collisions and less variability for peripheral collisions and a good way to quantitatively describe the variability change is to look at the variance of the collision probability density at various centralities. This variance is shown in Table (3.1). Note that the collision probability density for all the centrality classes has a mean of $x = 0.0$ fm, so most heavy quarks are produced at this point during a collision irrespective of the centrality.

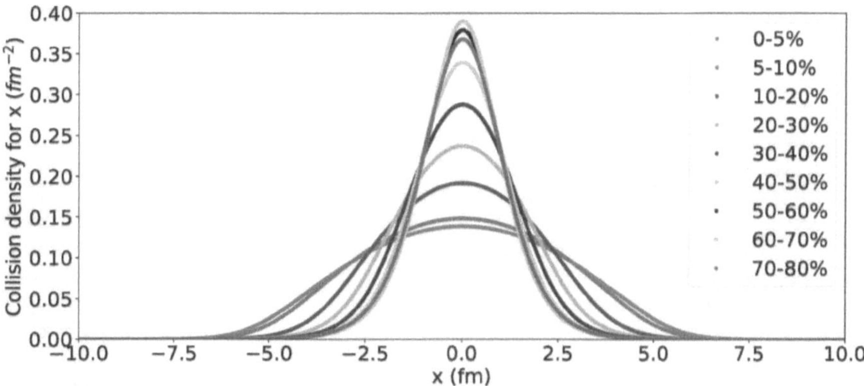

Figure 3.5: Collision probability density along the x direction for $Pb + Pb$ at $\sqrt{s_{NN}} = 5.5$ TeV for various centrality classes.

Table 3.1: Measure of variability for the collision probability density along x for various centrality classes of a $Pb + Pb$ collision at $\sqrt{s_{NN}} = 5.5$ TeV

Centrality	Variance
0-5%	0.0028
5-10%	0.0032
10-20%	0.0047
20-30%	0.0062
30-40%	0.0078
40-50%	0.0092
50-60%	0.010
60-70%	0.012
70-80%	0.0099

3.2 Heavy quarks produced in the transverse plane

When working with heavy-ion collisions, we need to produce heavy quarks in the transverse plane. This means that given the x coordinate, which we find using the procedure described in Sec. (3.1), we need to find the corresponding y coordinate. This requires the use of 2D interpolation schemes and these are discussed extensively in Ref. [120]. The procedure for generating random numbers on a plane is similar to the procedure we've discussed for producing random numbers along a line. The main differences are that; now we need to have a probability density function (PDF) which is a function of two variables, $f(x_1, x_2)$ and we need to compute the cumulative distribution function (CDF) in two dimensions, i.e. find the CDF(x_2) at every x_1 point.

The $PDF(x_1, x_2)$ is given by the collision probability density in Eq. (2.8) and describes the distribution of the heavy quark production points in the transverse plane. To find the CDF(x_2) at every x_1 point, we need to numerically compute the PDF in the form PDF(x_2) at every x_1 point. Note that the normalisation will be different for every x_1. The normalisation of the PDF(x_2) at each x_1 point is given by the integral of the collision probability density in Eq. (2.8) along x_2, at a fixed x_1. Notice that one can also find the CDF(x_1) at every x_2 point given the PDF(x_1) at every x_2 point. The choice is a matter of convenience but can be used as a consistency check since both approaches yield the same result.

Producing random numbers obeying the collision probability density on the transverse plane gives us the particle production points. We bin these points/coordinates to understand quantitatively where heavy quarks are produced. In Fig. (3.6) we show a 2D histogram of the production points for the $Pb + Pb$, 0-5% centrality class at $\sqrt{s_{NN}} = 5.5$ TeV. The colour-bar shows the number of heavy quarks present in a respective bin. We see that most heavy quarks are produced close to the origin (bins at and around the origin are densely populated) and the number of heavy quarks in each bin decreases as you move away from the origin.

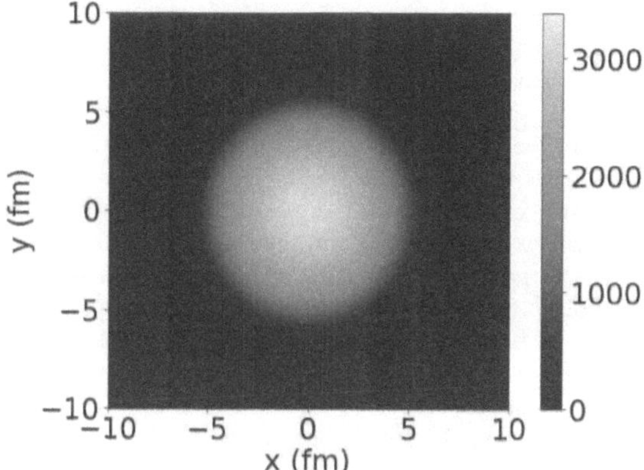

Figure 3.6: Unnormalised binned 2D collision probability density for the $Pb+Pb$ 0-5% centrality class at $\sqrt{s_{NN}} = 5.5$ TeV with a total of 2×10^6 Monte Carlo generated random numbers.

Recall that the impact parameter is in the reaction plane (along x) and the 0-5% centrality class corresponds to $\langle b \rangle = 2.37$ fm, so the collision is central. As a result, the binned points appear to be circles of different radii (depending on the number of heavy quarks present), this is what we'd expect to see for a central collision since almost all nucleons participate. In this case, the geometry of produced heavy quarks is almost symmetric around the origin and there's a fairly high number of heavy quarks produced at other points outside a radius of 2.5 fm from the origin.

In order to quantitatively compare this collision probability density of Monte Carlo generated random numbers to the 2D Glauber collision probability density given in Eq. (2.8), taking a ratio across the transverse plane is not sufficient. We need to take cross-sectional slices of this binned MC collision probability density at random points and compare the slices to corresponding slices in the collision probability density given by Eq. (2.8). Slices can either be taken along y at a particular point in x, or along x at a particular point in y. As an example, we'll take slices along y at $x = -4.95$ fm and $x = 0.05$ fm (these points correspond to the middle of bins and were picked randomly), this will give us a sense of how the collision probability density varies as we move across one coordinate.

In Fig. (3.7) we show a cross section of the binned 2D MC collision probability density at $x = -4.95$ fm compared to the Glauber y collision probability density at the same point in x. The corresponding ratio of the two cross-sections, which gives us a quantitative way of comparing our MC random number generation to the expected collision probability density as described by the Glauber model is shown in Fig. (3.8). The vertical bars included are the statistical uncertainties associated with the binning of the data as described by Poisson statistics. This ratio can be improved by increasing the number of heavy quarks produced (thus the number of heavy quarks in each bin).

To get a sense of how the collision probability density along y changes as we move across the x-direction, we take another cross sectional slice along y at $x = 0.05$ fm and this is shown in Fig. (3.9). Notice the change in the shape of the collision probability density, characterized by a change in the peak and extreme tails of the distribution (implying a change in variance as discussed earlier). At the point $x = 0.05$ fm, the distribution is wider and more heavy quarks are produced at points further away from $y = 0.0$ fm as compared to the slice taken at $x = -4.95$ fm where most heavy quarks were produced at and closely around $y = 0$ fm. Looking at cross sectional slices at points in $|x| \geq 5$ fm doesn't give us much information due to the low statistics (i.e. most of the bins have zero or less than five hundred heavy quarks in them) as can be seen from studying the colour-bar in Fig. (3.6).

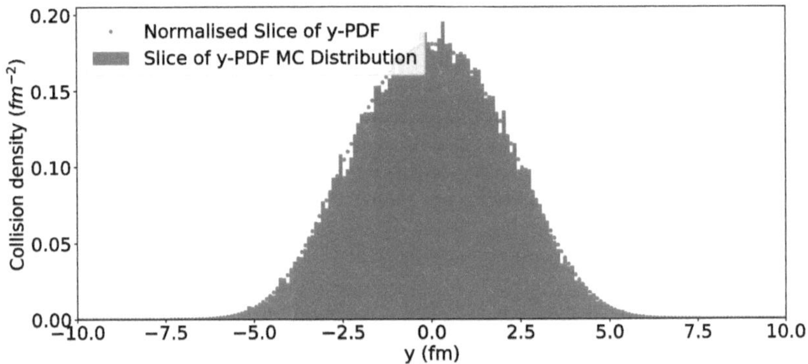

Figure 3.7: Cross section (along y) of the binned 2D collision probability density at $x = -4.95$ fm for the $Pb + Pb$ 0-5% centrality class at $\sqrt{s_{NN}} = 5.5$ TeV. The histogram shows Monte Carlo generated random numbers obeying this distribution.

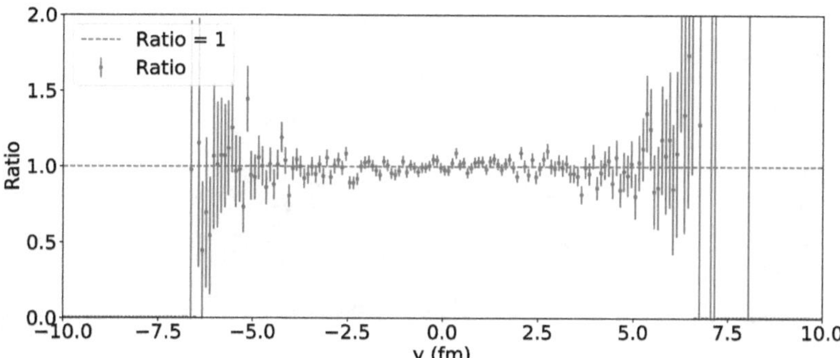

Figure 3.8: Ratio of the MC distribution cross section at $x = -4.95$ fm to the slice of the 2D collision probability density taken along y at $x = -4.95$ fm for the $Pb + Pb$ 0-5% centrality class at $\sqrt{s_{NN}} = 5.5$ TeV.

The ratio of the cross section shown in Fig. (3.9) is shown in Fig. (3.10). This ratio fluctuates closely around one for a wider range of y-values as compared to Fig. (3.8) because the distribution is more spread, with more heavy quarks in each of the bins on the tails of the distribution and containing fewer bins with zero heavy quarks in them. This ratio is quantitatively better than the one we saw at the point $x = -4.95$ fm because the bins along y at the point $x = 0.05$ fm have more heavy quarks in them as can be seen in Fig. (3.6), thus we have better statistics at this point.

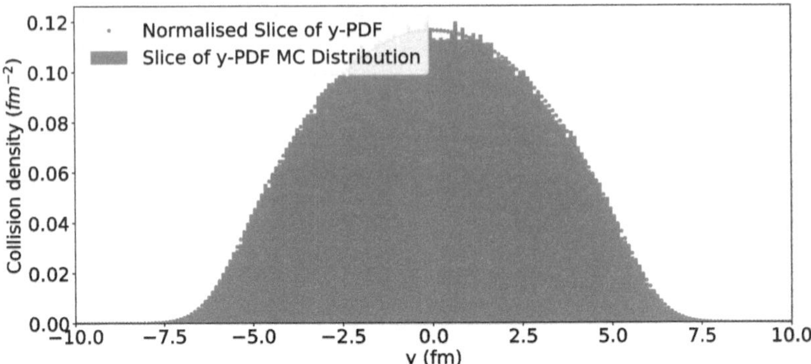

Figure 3.9: Cross section (along y) of the binned 2D collision probability density at $x = 0.05$ fm for the $Pb + Pb$ 0-5% centrality class at $\sqrt{s_{NN}} = 5.5$ TeV. The histogram shows Monte Carlo generated random numbers obeying this distribution.

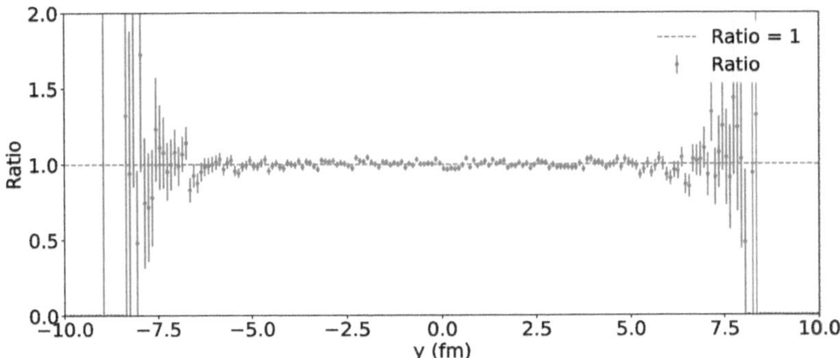

Figure 3.10: Ratio of the MC distribution cross section at $x = 0.05$ fm to the slice of the 2D collision probability density taken along y at $x = 0.05$ fm for the $Pb + Pb$ 0-5% centrality class at $\sqrt{s_{NN}} = 5.5$ TeV.

We can also perform this analysis for non-central collisions to get a sense of how the 2D collision probability density changes as we move from central to peripheral collisions. In Fig. (3.11) we show the unnormalised binned 2D collision probability density for the $Pb + Pb$ 20-30% centrality class at $\sqrt{s_{NN}} = 5.5$ TeV. The 20-30% centrality class corresponds to $\langle b \rangle = 7.92$

fm so the collision is semi-central. Notice the change in the shape of the binned 2D collision probability density from a circular shape in central collisions to an oval shape. This change in shape is representative of the overlap region since fewer nucleons participate in non-central collisions and most of the nucleons continue moving in their original direction (spectators). In this case, much more heavy quarks are produced at and close to the origin compared to the 0-5% centrality class and most of the other bins further from the origin have few or zero heavy quarks in them.

In Fig. (3.12) we show a sketch of the geometry of a non-central heavy ion collision, we see that the nuclear overlap region is elongated in the y direction (which is perpendicular to the reaction plane) and this is the same elongation depicted by the 2D binned collision probability density. For a broader discussion on the anisotropy due to this geometric shape of the overlap region, see Ref. [138]. The resulting $v_2(p_T)$ will be discussed in Chap. (5-6). As the collisions become more peripheral, the elongated overlap region becomes smaller due to fewer nucleon-nucleon collisions and most heavy quarks are produced at the origin and the immediate surrounding region.

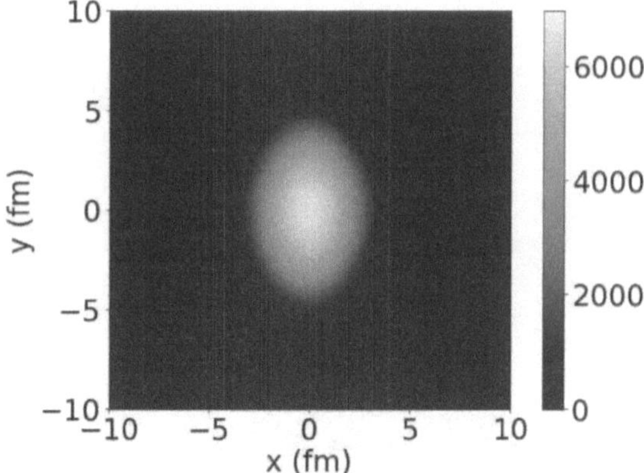

Figure 3.11: Unnormalised binned 2D collision probability density for the $Pb + Pb$ 20-30% centrality class at $\sqrt{s_{NN}} = 5.5$ TeV with a total of 2×10^6 Monte Carlo generated random numbers.

In Fig. (3.13) we show a cross sectional slice along y at the point $x = 0.05$ fm for the 20-30% centrality class. Notice the minimal change in the shape of the collision probability density which is due to the slices being taken along y (given that the overlap region is also elongated along y). The change in the collision probability density will be more pronounced for slices

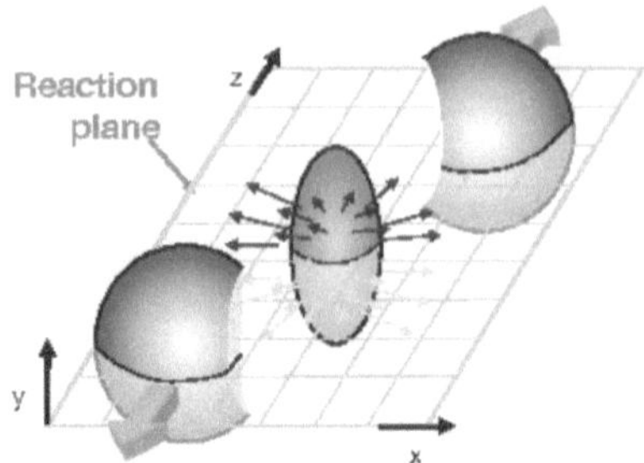

Figure 3.12: Geometry of a non-central heavy ion collision showing expanding QGP (centre) and spectator nucleons receding after the collision [138].

taken along x (which is parallel to the reaction plane). In Fig. (3.14) we show the corresponding ratio of this cross section to the slice of the Glauber collision probability density at $x = 0.05$ fm.

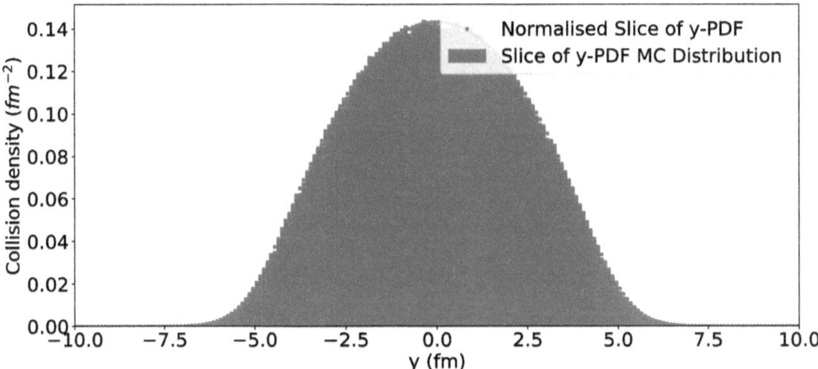

Figure 3.13: Cross section (along y) of the binned 2D collision probability density at $x = 0.05$ fm for the $Pb + Pb$ 20-30% centrality class at $\sqrt{s_{NN}} = 5.5$ TeV. The histogram shows Monte Carlo generated random numbers obeying this distribution.

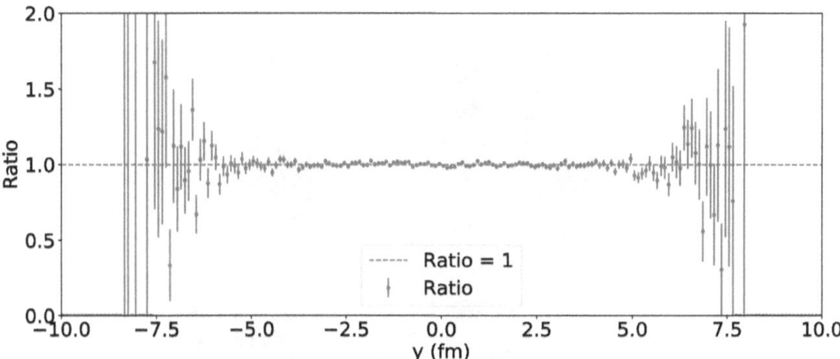

Figure 3.14: Ratio of the MC distribution cross section at $x = 0.05$ fm to the slice of the 2D collision probability density taken along y at $x = 0.05$ fm for the $Pb + Pb$ 20-30% centrality class at $\sqrt{s_{NN}} = 5.5$ TeV.

Lastly, Fig. (3.15) shows the unnormalised binned 2D collision probability density for the $Pb + Pb$ 70-80% centrality class at $\sqrt{s_{NN}} = 5.5$ TeV which corresponds to $\langle b \rangle = 13.75$ fm and is the most peripheral collision we'll study. As can be seen from the colour-bar, much more heavy quarks are produced at and around the origin as compared to the 20-30% centrality class. Thus concentrating the collision probability density around a small region of approximately $|x| < 2$ fm and $|y| < 3.5$ fm. The elongated overlap region becomes smaller since most nucleons just pass through without interacting (spectators).

For this centrality class, when taking a slice along y at the point $x = 0.05$ fm as shown in Fig. (3.16), it is easier to see a clear change in the shape of the collision probability density. The collision probability density has become less spread and strongly peaked at $y = 0$ fm and this is the same pattern we observed for peripheral collisions in the collision probability density along x (with y integrated out) shown in Fig. (3.5). We show the ratio of this cross sectional slice to the Glauber collision probability density for y at $x = 0.05$ fm with statistical uncertainties in Fig. (3.17). The ratio fluctuates around one and most points deviate from one by less than 5% (especially those in $|y| < 4$ fm due to the high statistics in that region). The overall result can be improved by increasing the number of quarks produced.

The collision probability density in Eq. (2.8) allows us to produce heavy quarks either along one axis or across the entire transverse plane in the event of a heavy-ion collision. It is easy to compare heavy quarks produced along a line (one axis) through binning them and comparing them to the collision probability density found by integrating out one of the coordinates in Eq.

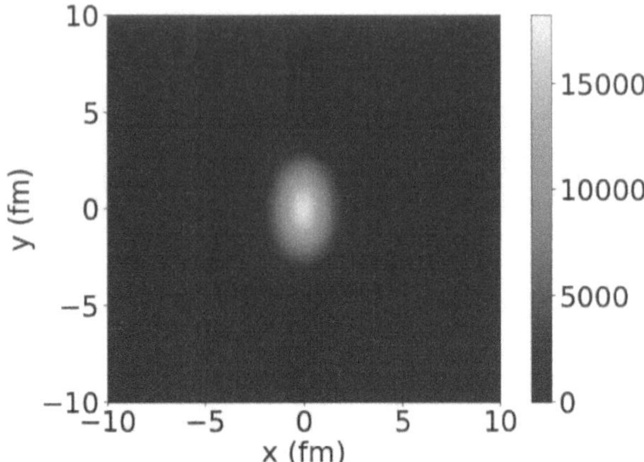

Figure 3.15: Unnormalised binned 2D collision probability density for the $Pb + Pb$ 70-80% centrality class at $\sqrt{s_{NN}} = 5.5$ TeV with a total of 2×10^6 Monte Carlo generated random numbers.

(2.8) then take the ratio at points corresponding to the centres of the peaks of the bins. The result can be improved by increasing the number of heavy quarks produced thus resulting in higher statistics. We also showed that as we move from central to peripheral collisions, the shape of the collision probability density changes from being more spread to being less spread and strongly peaked at $x = 0.0$ fm.

Heavy quarks produced in the transverse plane can also be binned in the xy-plane to get a qualitative sense of what the production geometry looks like. However, given the numerical sensitivity of the system that we're working with, it is important to have a quantitative analysis to test for the accuracy of our results. Taking the ratio of the 2D histogram of Monte Carlo generated random numbers to the collision probability density found by Eq. (2.8) is not good enough. So we take cross-sectional slices (along y) of the Monte Carlo data binned in the xy-plane at certain points in x, and compare these slices to the Glauber collision probability density the same points in x by taking the ratio. A good correspondence is found between the two densities computed using different methods as shown by the small size of statistical uncertainties in the ratio plots. Therefore, the Monte Carlo method of generating heavy quarks in the hydrodynamic medium is consistent with the Glauber model collision probability density.

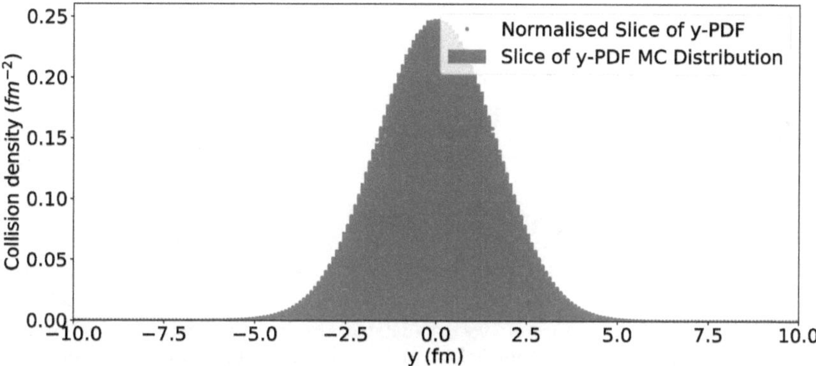

Figure 3.16: Cross section (along y) of the binned 2D collision probability density at $x = 0.05$ fm for the $Pb + Pb$ 70-80% centrality class at $\sqrt{s_{NN}} = 5.5$ TeV. The histogram shows Monte Carlo generated random numbers obeying this distribution.

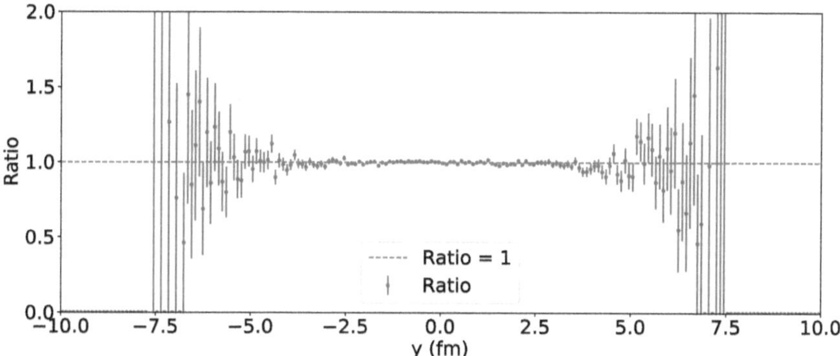

Figure 3.17: Ratio of the MC distribution cross section at $x = 0.05$ fm to the slice of the 2D collision probability density taken along y at $x = 0.05$ fm for the $Pb + Pb$ 70-80% centrality class at $\sqrt{s_{NN}} = 5.5$ TeV.

Chapter 4

Langevin Energy Loss

Once each heavy quark is produced in the geometry as discussed in Chap. (3), at thermalisation time, the QGP medium forms and the heavy quark propagates through it while interacting with the medium partons. The dynamics of a heavy quark propagating through a QGP medium are described by the Langevin equation given by Eq. (4.1) [88],

$$\frac{dp_i}{dt} \;=\; -\mu p_i + F_i^L + F_i^T \tag{4.1}$$

$$\mu \;=\; \frac{\pi\sqrt{\lambda}T^2}{2M_Q} \tag{4.2}$$

$$D \;=\; \frac{2T^2}{\kappa_L} \tag{4.3}$$

in the fluid's rest frame, where p^i is the three-momentum of an on-shell heavy quark that is moving with constant velocity in the plasma, μ is the drag loss coefficient of a heavy quark [90] (i.e. the momentum loss per unit Δt), M_Q is the mass of the heavy quark in a plasma of temperature T, λ is the 't Hooft coupling constant and D is the diffusion constant. The stochastic forces (diffusion terms) F_i^L and F_i^T are the longitudinal and transverse momentum kicks with respect to the quark's direction of propagation. The random fluctuating momentum kicks satisfy the following relations;

$$\langle F_i^L(t_1)F_j^L(t_2)\rangle \;=\; \kappa_L \hat{p}_i\hat{p}_j g(t_2 - t_1) \tag{4.4}$$

$$\langle F_i^T(t_1)F_j^T(t_2)\rangle \;=\; \kappa_T(\delta_{ij} - \hat{p}_i\hat{p}_j)g(t_2 - t_1) \tag{4.5}$$

where $\hat{p}_i = p_i/|\vec{p}|$ is the unit 3-vector in the direction of the momentum, g is a function only known numerically [88] and the coefficients κ_L and κ_T are the respective variances of the transverse and longitudinal momentum transfers per unit Δt and are given by,

$$\kappa_T \;=\; \pi\sqrt{\lambda}T^3\gamma^{1/2} \tag{4.6}$$

$$\kappa_L \;=\; \gamma^2\kappa_T = \pi\sqrt{\lambda}T^3\gamma^{5/2} \tag{4.7}$$

$$\gamma = \frac{1}{\sqrt{1-v^2}} \tag{4.8}$$

$$\hat{q} = \langle p_\perp(t)^2 \rangle / L \approx \gamma(2\pi T^3 \sqrt{\lambda})/v \tag{4.9}$$

where γ is the Lorentz factor for the heavy quark (with velocity v), $\hat{q}$ is the jet-quenching parameter and $\langle p_\perp(t)^2 \rangle$ is the average transverse momentum that the heavy quark acquires after it has traveled a distance L. This construction of parameters does not obey the fluctuation-dissipation theorem [88] and the transport scheme only leads to thermalization in the $p_T \to 0$ limit where the fluctuation-dissipation theorem is satisfied. Energy loss computations performed using this set of parameters will be labelled **D(p)** since the diffusion coefficient depends on the heavy quark's momentum.

The longitudinal direction of the heavy quark is the most important direction for calculations of suppression observables. As shown in Eq. (4.7), the momentum fluctuations in the longitudinal direction grow very quickly (i.e. as $\gamma^{5/2}$) with the velocity of the heavy quark. As a result, with increasing velocity (for high momentum), the energy loss model employing this set of parameters breaks down as these momentum fluctuations significantly impact the energy loss of the heavy quark. A speed limit where this happens is estimated by restricting the momentum that is picked up by the heavy quark through momentum fluctuations over the time scale set by the drag coefficient to a value that is small when compared to the total momentum of the heavy quark, and is given in Eq. (4.10) [88].

$$\gamma \lesssim \gamma_{crit}^{fluc} = \frac{M_Q^2}{4T^2} \tag{4.10}$$

There are various approaches used to limit the momentum growth of the heavy quark and one is discussed in Appx. (C.2). A different set of parameters is obtained by requiring the momentum fluctuations to obey the fluctuation-dissipation theorem. The diffusion and drag coefficients are related by the Einstein relation as shown in Eq. (4.11) [91],

$$\mu = \frac{\kappa}{2MT} \tag{4.11}$$

where $\kappa_T = \kappa_L = \kappa$, while the diffusion constant, D, is related to the drag loss coefficient by Eq. (4.12) [89].

$$D = \frac{T}{M\mu} = \frac{2T^2}{\kappa} = \frac{2}{\pi\sqrt{\lambda}T} \tag{4.12}$$

The momentum fluctuations are given by Eq. (4.13) [91],

$$\kappa = \pi\sqrt{\lambda}T^3 \tag{4.13}$$

while the jet quenching parameter $\hat{q}$ is given by Eq. (4.14) [139].

$$\hat{q} = \langle p_\perp(t)^2 \rangle / L \approx (2\pi T^3 \sqrt{\lambda})/v \tag{4.14}$$

Computations using these parameters will be labelled **D=const** since the diffusion coefficient does not depend on the heavy quark's momentum/velocity.

The drag loss coefficient, μ has a temperature and 't Hooft coupling dependence and needs to be mapped from $\mathcal{N} = 4$ SYM theory to QCD. Two sets of parameters are used to account for the systematic theoretical uncertainties associated with the mapping of a result in $\mathcal{N} = 4$ SYM theory to QCD [88, 89]. These parameters are as follows:

1. **Reasonable (Re):**

$$T_{SYM} = T_{QCD}, \quad \lambda = 4\pi \alpha_s N_c = 4\pi \times 0.3 \times 3 \simeq 11.3 \tag{4.15}$$

2. **Gubser (Gb):**

$$T_{SYM} = 3^{-1/4} T_{QCD}, \quad \lambda = 5.5 \tag{4.16}$$

These sets of parameters are discussed in more detail in Appx. (C.1). The Reasonable parameters compare QCD to $\mathcal{N} = 4$ SYM theory at the same temperature. However, this doesn't give a good overlap of the static force between a quark and antiquark, $\alpha_{q\bar{q}}(r, T)$ and $\alpha_{SYM}(r, T)$ between the two theories [140]. Also, comparing the two theories at fixed temperature results in overestimates of quantities such as drag effects and the screening length in QCD, which is partially corrected by comparing the two theories at fixed energy [140]. The 't Hooft coupling is fixed by equating the coupling in $\mathcal{N} = 4$ SYM and the coupling in the QCD Lagrangian, $g_{YM} = g_s$. The Gubser framework is based on the relation of the energy densities between the two plasmas (i.e. QCD and $\mathcal{N} = 4$ SYM plasma), in that, the energy density, $\epsilon \propto T^4$ in $\mathcal{N} = 4$ SYM while ϵ is approximately proportional to T^4 in QCD for $T \gtrsim 1.1 T_c$ [140] and requiring the condition $\epsilon_{SYM} = \epsilon_{QCD}$ to hold results in the temperature relation given by Eq. (4.16). In addition, the 't Hooft's coupling in $\mathcal{N} = 4$ SYM is normalised by comparing $\alpha_{q\bar{q}}(r, T)$ and $\alpha_{SYM}(r, T)$ which yields $\lambda = 5.5$ [140].

4.1 Description of the energy loss model

The work of this **book** is based on the energy loss model developed in [88] and expanded on in [113, 139]. In detail, and as described in [88], in the D(p) scenario, the Stratonovich stochastic differential equation implemented as an Itô SDE in the Euler-Maruyama scheme is

$$p_{n+1}^{\prime i} = \left[1 - \mu \, dt' + \frac{1}{2}\kappa \, dt'\left(\frac{5\gamma^{5/2}}{4E'^2} + \frac{(d-1)\,\gamma^{1/2}}{(\gamma^2 + 1)\,M_Q^2}\right)\right]p_n^{\prime i} + C^{ij}dW_j \tag{4.17}$$

$$\mu = \frac{\pi\sqrt{\lambda}T^2}{2M_Q} \tag{4.18}$$

$$C^{ij} = \sqrt{dt'}\,\kappa\gamma^{1/4}\left(\frac{p'^i p'^j}{(\gamma^2+1)\,M_Q^2} + \delta^{ij}\right). \tag{4.19}$$

where M_Q is the mass of the heavy quark, dt' is the time step dt boosted into the local rest frame of the fluid; $dt' = dt/\gamma$; $\kappa = \pi\sqrt{\lambda}T^3$, where T is the temperature of the fluid in its local rest frame; d is the number of spatial dimensions in the calculation. The number of spatial dimensions, d, is informed by our hydrodynamics backgrounds which the heavy quarks propagate through. These backgrounds (for the medium evolution) are generated by VISHNU 2+1D viscous relativistic hydrodynamics [115, 116]. The dW_j are the uncorrelated, Gaussian Wiener kicks with mean zero and standard deviation one. Note that heavy quarks are the fundamental degrees of freedom in our energy-loss approach, as compared to fluid cells in hydrodynamics, however, we discuss hydrodynamics in Appx. (B) because the heavy quarks propagate through a medium that is based on hydrodynamics.

In detail, in the D=const scenario, the Itô SDE in the Euler-Maruyama scheme we implement is

$$p'^i_{n+1} = \left(1 - \mu\,dt'\right)p'^i_n + C^{ij}dW_j \tag{4.20}$$

$$\mu = \frac{\pi\sqrt{\lambda}T^2}{2E'} \tag{4.21}$$

$$C^{ij} = \sqrt{2dt'\,\mu E'\,T}\,\delta^{ij}. \tag{4.22}$$

In both the D(p) and D=const scenarios, the heavy quark is propagated in coordinate space according to

$$x^i_{n+1} = \frac{p^i_{n+1}}{E_{n+1}}dt, \tag{4.23}$$

dt was taken to be $1/150 \times \mu_{max}$, where μ_{max} is the drag coefficient at the center of the fireball at the thermalization time, the largest drag coefficient for any individual collision [88].
where μ_{max} gives the drag coefficient at the centre of the QGP medium at thermalisation time, this is the largest drag coefficient for any individual collision. The value of dt is obtained through a trial and error in order to consistently yield stable static thermal distribution in Ref. [88]. In computing the trajectory of the heavy quark from its production point, the energy loss code used boosts the heavy quark into the fluid's local rest frame at each time step, evaluates the change in momentum then boosts back to the lab frame [88].

The momentum production spectrum of the heavy quarks is obtained from FONLL [96–98] and is shown in Fig. (4.1). The spectrum peaks at $p_T \sim 3$ GeV/c and has a power law dependence on momentum. The initial direction of propagation of the heavy quarks is randomly sampled following a uniform distribution. Starting from the point of production, at thermalisation, these heavy quarks propagate through the QGP medium until a certain time has lapsed where hadronisation starts to occur (i.e. the maximum time of the VISHNU background) or until the temperature of the medium drops below a critical temperature (T_c). The pseudo-random number generation process was performed using the *Ran* routine discussed in Ref. [120] and is discussed in Chap. (3). The seed generation follows the routine described in Ref. [141].

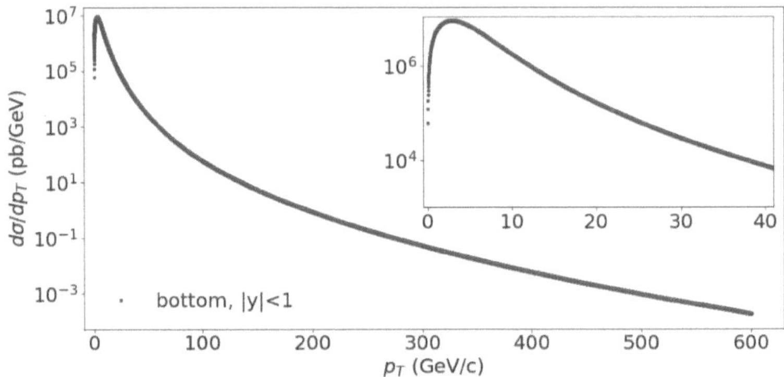

Figure 4.1: Bottom quark production in pp collisions at the LHC at $\sqrt{s_{NN}} = 5.5$ TeV for $|y| < 1$, the y-axis is in log-scale.

The position as well as the momentum of a single bottom quark propagating through a VISHNU background, obtained from our model with parameters **Gb, D(p)** is shown in Fig. (4.2-4.5). Note that D(p) parameters are string theory predictions (hence the Einstein relations are not imposed) and momentum fluctuations in the longitudinal direction grow very quickly (i.e. as $\gamma^{5/2}$) with the velocity of the heavy quark. The gap from $t = 0$ fm/c to $t = 0.6$ fm/c is the time between the heavy quark production and thermalisation. Various centrality classes are compared for a quark starting at the same position with the same initial momentum. The bottom quark starts with a high momentum and loses some of its momentum as it propagates through the medium, while its direction barely changes. Notice in Fig. (4.2) that the time spent by the bottom quark in the QGP medium decreases as we move from central to peripheral collisions. The reason for the difference in the time spent by the heavy quark in the medium is that more QGP is produced in central collisions compared to peripheral collisions.

In Fig. (4.3) we show the position and momentum of a bottom quark produced at the same point as the quark in Fig. (4.2), but with a high initial momentum. In addition to the time

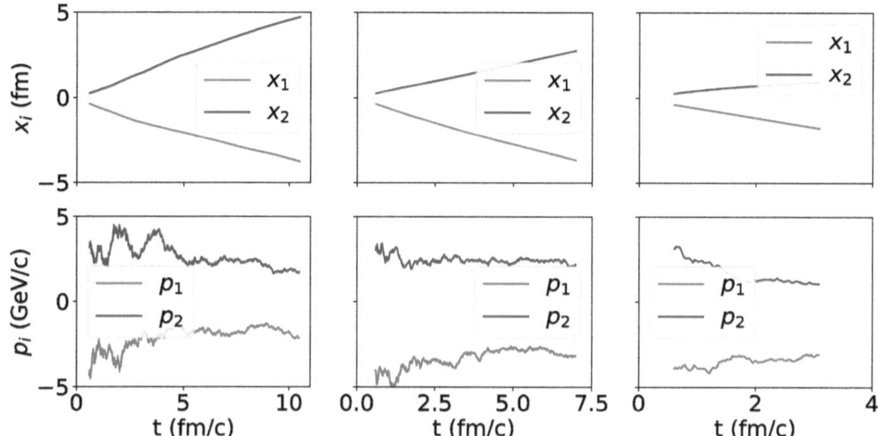

Figure 4.2: Position and Momentum of a single bottom quark produced at (0,0) fm with initial momentum (-4,3) GeV/c propagating through a VISHNU [115, 116] hydrodynamic background for different centralities as follows: 0-5% (Left), 30-40% (Middle) and 70-80% (Right).

dependence on centrality discussed in the previous paragraph, we see that the time that the bottom quark spends in the QGP medium has decreased for each centrality class. This decrease in the time is because the quark starts with a high momentum so it propagates through the medium and exits quicker than a quark starting off with intermediate or low momentum. Also, the bottom quark loses more energy compared to the quark starting with intermediate momentum. Looking at the bottom panel of Fig. (4.3), we see that due to the amount of QGP the heavy quark needs to propagate through, the energy loss decreases as a function of centrality as we move from central to semi-central and ultimately peripheral collisions.

If instead, the bottom quark is produced at a region away from the origin with intermediate initial momentum, its direction barely changes and it doesn't lose much of its momentum in the time frame that it spends propagating through the QGP medium. This case is illustrated in Fig. (4.4) which shows the position and momentum of a bottom quark as a function of time in the transverse plane. When compared to the heavy quark produced at the origin with the same initial momentum shown in Fig. (4.2), the heavy quark now spends less time (less than half the time for the 30-40% centrality class) in the medium and also travels a shorter distance. The bottom quark illustrated in Fig. (4.5) is produced at the same point as the quark in Fig. (4.4), but with a high initial momentum. This quark spends less time in the medium since it starts with a higher momentum. However, when compared to a heavy quark produced with the same initial momentum but at the origin, shown in Fig. (4.3), it loses less energy since there's less medium to propagate through.

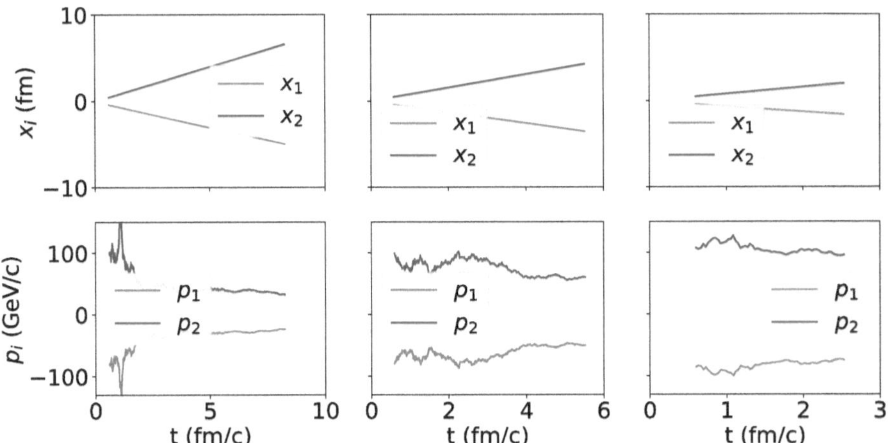

Figure 4.3: Position and Momentum of a bottom quark produced at (0,0) fm with initial momentum (-80,100) GeV/c propagating through a VISHNU [115, 116] hydrodynamic background for different centralities as follows: 0-5% (Left), 30-40% (Middle) and 70-80% (Right).

Note from Fig. (4.4) and Fig. (4.5) that we've excluded the 70-80% centrality class for this comparison because the medium is cold at the production location under consideration for this centrality class, thus cannot be regarded as QGP. Some conclusions to draw from this discussion are; low momentum quarks lose less momentum when compared to high momentum quarks, this is more pronounced in peripheral collisions, low momentum quarks propagate through the medium until a time where hadronisation begins (informed by the VISHNU background), while high momentum quarks generally propagate through the medium until the temperature of the medium drops below a critical temperature and heavy quarks produced in peripheral collisions spend less time in the medium compared to those produced in central collisions. Note that the discussion on the time the heavy quarks spend in the medium will be crucial when we discuss $v_2(p_T)$ results in Chap. (6.2) since v_2 is picked up from propagating through different lengths of material and is generally higher in the intermediate momentum region compared to high momentum.

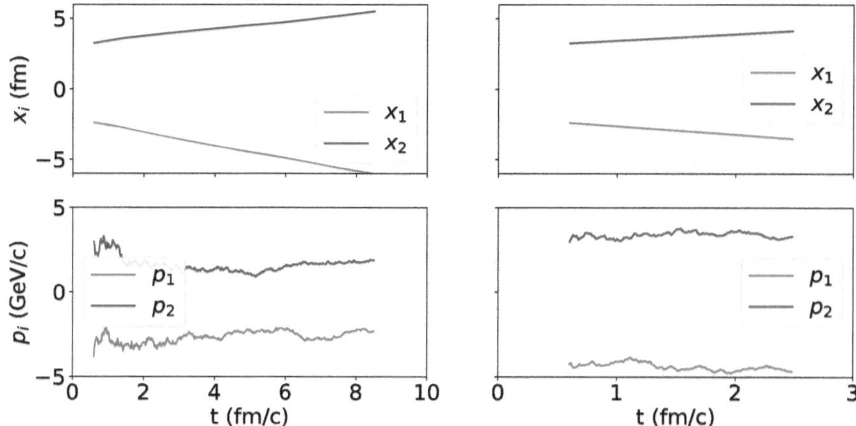

Figure 4.4: Position and Momentum of a single bottom quark produced at (-2,3) fm with initial momentum (-4,3) GeV/c propagating through a VISHNU [115, 116] hydrodynamic background for different centralities as follows: 0-5% (Left) and 30-40% (Right).

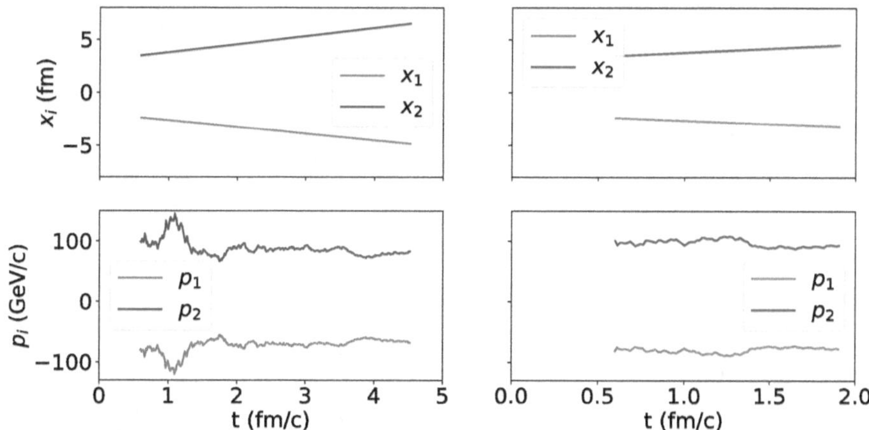

Figure 4.5: Position and Momentum of a single bottom quark produced at (-2,3) fm with initial momentum (-80,100) GeV/c propagating through a VISHNU [115, 116] hydrodynamic background for different centralities as follows: 0-5% (Left) and 30-40% (Right).

Chapter 5

Measured Observables

5.1 Computing $R_{AA}(p_T)$

The nuclear modification factor, $R_{AA}(p_T)$, provides us with a possible approach to quantify the difference between nucleus-nucleus (AA) collisions and nucleon-nucleon (pp) collisions. In simple terms, it is a ratio of the number of quarks (or final state particles) with a particular p_T in AA collisions to the number of quarks (or final state particles) in pp collisions, scaled by the respective cross sections. The $R_{AA}(p_T)$ is given by Eq. (5.1),

$$R_{AA}(p_T) = \frac{\sigma_{tot}^{pp} dN^{AA}/dp_T}{N^{AA} d\sigma^{pp}/dp_T} \tag{5.1}$$

$$\sigma_{tot}^{pp} = \int_{p_{Tmin}}^{p_{Tmax}} \frac{d\sigma^{pp}}{dp_T} dp_T \tag{5.2}$$

$$N^{AA} = \int_{p_{Tmin}}^{p_{Tmax}} \frac{dN^{AA}}{dp_T} dp_T \tag{5.3}$$

where dN^{AA}/dp_T is the meson spectrum in AA collisions and $d\sigma^{pp}/dp_T$ is the meson spectrum in pp collisions.

Note that there's a difference in which we've defined $R_{AA}(p_T)$ in Eq. (5.1 - 5.3) compared to the standard definition used in experiment given in Eq. (1.3 - 1.4). However, one can easily show that in both these cases, $R_{AA}(p_T)$ is dimensionless. Experimentally, there is a trivial difference in the number of quarks (i.e. b-quarks) and consequently mesons produced in pp collisions compared to those produced in AA collisions. In pp collisions, we have two nucleons colliding, while in AA collisions (i.e. central $Pb + Pb$ collision), we have $\sim 208 + 208$ nucleons colliding, thus producing more quarks (consequently more mesons). The way that one scales out this difference in the number of produced particles (due to more nucleons colliding), is by normalising by N_{coll} as shown in Eq. (1.3).

On the other hand, the AA spectrum (dN^{AA}/dp_T) in Eq. (5.1) doesn't include the enhancement in the number of quarks (consequently mesons) produced due to the number of binary collisions. The reason why there is no enhancement in the number of quarks produced is that theoretically, we embed a pp spectrum inside of an AA collision without scaling up this pp spectrum by the number of binary collisions while also ensuring that the number of quarks (consequently mesons) is conserved. Hence we normalise by the ratio of the total number of particles in pp to the total number of particles in AA collisions as shown in Eq. (5.1).

Fragmentation

Approximately 10 fm/c [133] after a heavy-ion collision, hadronisation starts to occur from the QGP. Given that this is a very quick process, we have no direct way of measuring heavy quarks in the QGP experimentally. Detectors can only measure final state particles (i.e. hadrons), however, the process of hadronisation is still challenging in quantum chromodynamics (QCD). QCD bound-states are non-perturbative in nature and there is no first principle approach to describe their formation. In the energy-loss code, the heavy quarks fragment into their final states and we'll discuss the fragmentation process in this section.

Perturbative quantum chromodynamics (pQCD) can be used to describe inclusive hadron production at sufficiently large momentum transfer. We assume that the heavy quarks fragment in vacuum, and hence we may employ the FONLL fragmentation functions (D_{NP}) [142]. This function accounts for all low-energy contributions, such as the process where a heavy quark turns into a heavy-flavoured hadron and gives us the probability that a parton will fragment into a hadron and is given by Eq. (5.4).

$$D_{NP}(z) \;=\; Norm. \times \frac{1}{1+c}[\delta(1-z) + cN_{a,b}^{-1}(1-z)^a z^b], \quad N_{a,b} = \int_0^1 (1-z)^a z^b dz \quad (5.4)$$

In this **book**, for bottom quarks fragmenting into B mesons, we have used the Kartvelishvili et al. distribution [143] given by Eq. (5.5),

$$D_{NP}^{b \to B}(z) \;=\; (\alpha+1)(\alpha+2)z^\alpha(1-z), \quad z = \frac{p_{meson}}{p_{quark}}, \quad z \in [0,1] \quad (5.5)$$

where $\alpha = 24.2$, is the fragmentation parameter obtained in Ref. [142] corresponding to $m_b = 4.75$ GeV, which is the central value for the mass of bottom quarks. However, these two fragmentation functions yield the same result and can be used to check for consistency. The Kartvelishvili et al. distribution is normalised to unity and it adopts the assumption that each bottom quark produces a b-hadron (same as $\bar{b}$). These non-perturbative fragmentation functions are universal, in that, they are measured in e^+e^- annihilation processes and are applicable to the description of hadron production in other hard QCD processes. Parameters such as α are extracted from finding fits to fragmentation data from various experiments [142]. In

Fig. (5.1) we show the fragmentation function given by Eq. (5.5).

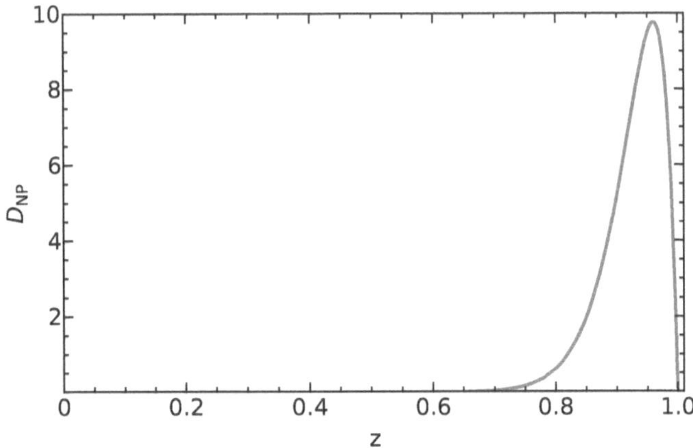

Figure 5.1: The Kartvelishvili et al. [143] non-perturbative distribution function given by Eq. (5.5)

Given a bottom quark spectrum, $dN^q/dp^q(p^q)$, either from AA or pp collisions, where p^q is the transverse momentum of the bottom quark, we can use the distribution discussed earlier to compute the B-meson spectrum, $dN^B/dp^B(p^B)$, that these bottom quarks decay to. Here, p^B is the transverse momentum of the B-mesons. This conversion is given by:

$$\frac{dN^B}{dp^B}(p^B) \;=\; \int dp^q dz \frac{dN^q}{dp^q}(p^q)\delta(p^B - zp^q)D_{NP}^{q\to B}(z) \tag{5.6}$$

$$=\; \int_0^1 \frac{dz}{z}\frac{dN^q}{dp^q}(p^B/z)D_{NP}^{b\to B}(z) \tag{5.7}$$

The heavy quark production spectrum falls sharply with transverse momentum (p_T), so most partons have small momentum as shown in Fig. (4.1) for bottom quarks at $\sqrt{s_{NN}} = 5.5$ TeV. In order to produce a hadron that has a momentum p_h by fragmentation, one needs the fragmenting parton to have a momentum $p_T/z > p_h$ [144]. However, fragmentation functions are highly weighted towards high values of z as shown in Fig. (5.1).

5.1.1 $R_{AA}(p_T)$ statistical uncertainties

The spectrum of produced particles from our energy loss code is given by $d^2N/dp_Td\phi$, where the azimuthal angle, ϕ in the data consists of six bins in the interval $0 < \phi < \pi/2$ and the statistical uncertainties in each $d^2N/dp_Td\phi$ bin are computed by employing Poisson statistics as per the following equation;

$$\sigma_i = \sqrt{n_i\left(1 - \frac{n_i}{N}\right)} \tag{5.8}$$

where n is the total number of quarks in each bin and N is the total number of quarks in the system averaged over the number of ϕ bins. These uncertainties are used to split each ϕ bin into three bins where one bin consists of the original $R_{AA}(\phi)$ spectra, the other bin consists of the original spectra plus the uncertainties $(dN/d\phi + \sigma)$ and the last bin consists of the original spectra less the uncertainties $(dN/d\phi - \sigma)$. Using the new bins, one can compute $R_{AA}(p_T, \phi)$ and the corresponding uncertainty is then computed by taking the square-root of the maximum of the difference of squares. The average $R_{AA}(p_T)$ is found by taking the average over all ϕ bins and the corresponding uncertainty is computed as follows [137];

$$\sigma_{avg} = \sqrt{\Sigma_{i=1}^{j}(\sigma_i)^2} \tag{5.9}$$

where j is the total number of ϕ bins and σ_i is the uncertainty in each ϕ bin for a fixed p_T.

5.2 Computing the $v_2(p_T)$

The $v_2(p_T)$ characterises the ellipticity (degree of deviation from circularity) of the azimuthal distribution of the produced particles in a heavy-ion collision, given that the distribution of final state particles is asymmetric in the azimuthal plane. The azimuthal asymmetry in the distribution of final state particles is a consequence of the heavy quarks propagating through different medium. The difference in the medium has a dependence on the centrality class and can be understood by looking at the geometry of the medium shown in Fig. (3.6) and comparing it to that shown in Fig. (3.15). Heavy quarks can still propagate through different medium even if they are produced in the same centrality class and this is a result of the different angles in which the heavy quarks are produced.

In this book, the $v_2(p_T)$ is computed by taking the Fourier expansion of the nuclear modification factor, $R_{AA}(p_T, \phi)$ obtained using the energy-loss code. The Fourier expansion of $R_{AA}(p_T, \phi)$ is given by Eq. (5.10).

$$R_{AA}(p_T, \phi) = R_{AA}(p_T)\left[1 + 2v_2(p_T)\cos(2\phi)\right] \tag{5.10}$$

where our $R_{AA}(p_T, \phi)$ result covers the transverse momentum domain, $0.25 < p_T < 599.75$ GeV/c and azimuthal angle bins $0 < \phi < \pi/2$ rad, which we extended to $0 < \phi < 2\pi$ rad by symmetry. The $v_2(p_T)$ is then extracted using the FindFit function in Mathematica [145]. As a consistency check, the $R_{AA}(p_T)$ result can also be extracted from this fitting procedure and compared to the result obtained by averaging $R_{AA}(p_T, \phi)$ over all ϕ bins.

One can also take the Fourier expansion of the azimuthal distribution of produced particles given in Eq. (5.11),

$$\frac{dN}{d\phi} = \frac{N}{2\pi}\left(1 + 2\sum_{n=1}^{\infty} v_n(p_T, y)\cos[n(\phi - \Psi_R)]\right) \tag{5.11}$$

where ϕ is the azimuthal angle of the detected particles and Ψ_R is the plane of the symmetry of the initial collision zone as shown in Fig. (D.1). The $dN/d\phi$ is the original spectra under consideration, i.e. the bottom quark spectra obtained in the energy loss model or the B-meson spectra obtained by fragmentation. This $dN/d\phi$ spectra should be a uniform distribution in the event where there is no azimuthal anisotropy. We can compute the $v_2(p_T)$ by taking the second Fourier coefficient of the $dN/d\phi$ spectra as shown in Eq. (5.12). [68].

$$v_2 = \frac{\int d\phi \frac{dN}{d\phi} \cos(2\phi)}{\int d\phi \frac{dN}{d\phi}} \tag{5.12}$$

The result in Eq. (5.12) gives us an alternative approach for computing $v_2(p_T)$ and we've used this approach as a consistency check for our $v_2(p_T)$ results.

5.2.1 $v_2(p_T)$ statistical uncertainties

Propagating statistical uncertainties for our $v_2(p_T)$ extracted from the Mathematica FindFit procedure is highly non-trivial. Given that there are nonlinear combinations of various quantities contributing to values in the $R_{AA}(p_T, \phi)$ bins, computing the statistical uncertainties analytically requires one to understand how derivatives of these quantities work. A less taxing approach that has been employed in this paper involves the use on Monte Carlo techniques [135–137].

Using the $R_{AA}(p_T, \phi)$ data and the corresponding uncertainties obtained by employing Poisson statistics as discussed in Sec. (5.1.1), one can create a random number generator that generates random sets of data based on the $R_{AA}(\phi)$ data for a specific value of p_T. The random number generator follows a Gaussian distribution, with the mean given by the value in each $R_{AA}(p_T, \phi)$ bin and the standard deviation given by the corresponding uncertainty in that bin. For each run of the random number generator, one then computes $v_2(p_T)$ using any preferred method

(Mathematica FindFit in this case). This can be repeated many times for each p_T (i.e. run the random number generator many times), computing $v_2(p_T)$ each time. Using this huge set of $v_2(p_T)$ data, one can compute the mean value of $v_2(p_T)$ and the corresponding standard deviation, which gives the uncertainty in $v_2(p_T)$.

As a consistency check, one can pick a particular $R_{AA}(p_T, \phi)$ bin and compare the mean and standard deviation of the Gaussian distribution to the corresponding values in the original $R_{AA}(p_T, \phi)$ data. This comparison can be used to inform the number of times in which one needs to run the random number generator for better accuracy.

Chapter 6

Results

In this book, we have assumed a strongly coupled plasma and have used the Langevin energy loss model to compute the suppression of heavy quarks as they propagate through the QGP medium. We will now discuss our main results. These results include the nuclear modification factor, $R_{AA}(p_T)$ and the $v_2(p_T)$ for B-mesons at $\sqrt{s_{NN}} = 2.76$ TeV, $|y| < 0.5$ for central $Pb + Pb$ collisions and $\sqrt{s_{NN}} = 5.5$ TeV, $|y| < 1$ for $Pb + Pb$ collisions at various centralities. The results will include a comparison of models employing "Gubser" parameters (labelled **Gb**) with $\lambda = 5.5$ and $T_{SYM} = 3^{-1/4}T_{QCD}$ as well as "Reasonable" parameters (labelled **Re**) with $\lambda = 4\pi \times 0.3 \times 3 \sim 11.3$ and $T_{SYM} = T_{QCD}$, discussed in Chap. (4). This comparison of models employing these parameters seeks to account for the systematic theoretical uncertainties associated with the mapping of parameters in $\mathcal{N} = 4$ SYM theory to QCD. The results will also include a comparison of a model where the diffusion coefficient is a function of the heavy quark's momentum, labelled $\mathbf{D} = \mathbf{D(p)}$ and a model where the diffusion coefficient does not depend on momentum, labelled $\mathbf{D} = \mathbf{const}$, discussed in Chap. (4). We use these two limits to try and account for the uncertainty related to how the diffusion coefficient behaves as a function of momentum in AdS/CFT.

6.1 $R_{AA}(p_T)$

In Fig. (6.1) we show the predictions for the nuclear modification factor, $R_{AA}(p_T)$ for B-mesons at $\sqrt{s_{NN}} = 2.76$ TeV at $0 - 10\%$ centrality in the rapidity range $|y| < 0.5$. The vertical bars represent statistical uncertainties computed using the method described in Chap. (5.1.1). The size of the statistical uncertainties grow with transverse momentum as there are fewer bottom quarks and consequently B-mesons in high transverse momentum bins. As a result of energy loss, most high transverse momentum B-mesons lose their momentum as they traverse through the QGP medium and end up in low momentum bins. A value of $R_{AA}(p_T) < 1$ indicates that bottom quarks (and consequently the B-mesons that they decay to) produced with initial momentum p_T, are suppressed as they propagate through the quark-gluon plasma. This signals the presence of a medium (QGP) produced during these heavy-ion collisions, that has an effect on the particles that propagate through it. If instead, the $R_{AA}(p_T) = 1$, then there is no energy loss, which indicates that either there is no medium produced (as is the case in pp collisions) or the medium produced does not induce effects on the particles propagating through it that can be observed in their final state.

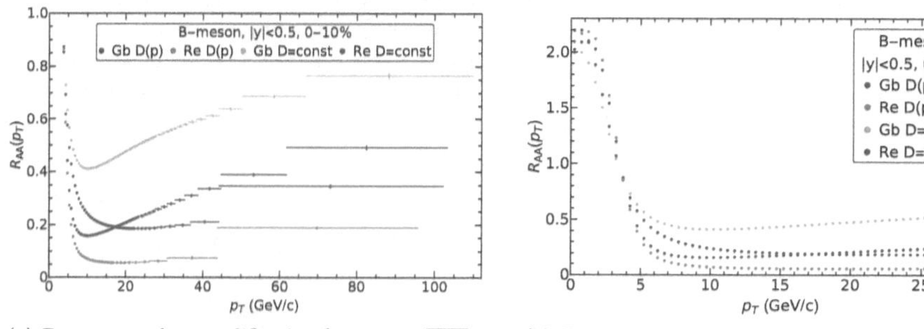

(a) B-meson nuclear modification factor at $\sqrt{s_{NN}} = 2.76$ TeV for the 0-10% centrality class.

(b) Expanded view of the transverse momentum region, $0 < p_T \leq 30$ GeV/c of Fig. (a), on the left.

Figure 6.1: B-meson nuclear modification factor at $\sqrt{s_{NN}} = 2.76$ TeV for Gb and Re parameters with a constant diffusion coefficient and one that is dependent on the momentum.

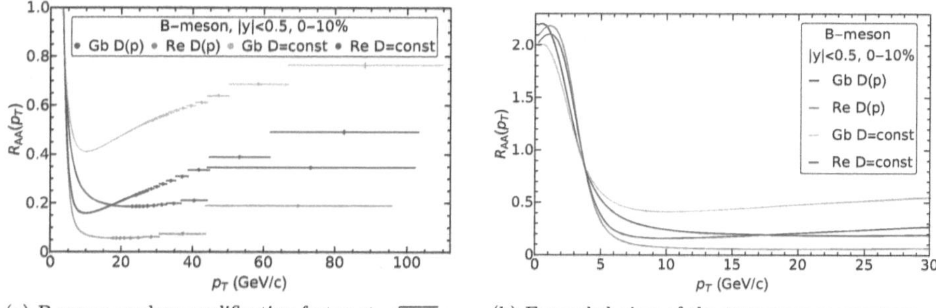

(a) B-meson nuclear modification factor at $\sqrt{s_{NN}} = 2.76$ TeV for the 0-10% centrality class.

(b) Expanded view of the transverse momentum region, $0 < p_T \leq 30$ GeV/c of Fig. (a), on the left, including the region $R_{AA}(p_T) > 1$.

Figure 6.2: B-meson nuclear modification factor at $\sqrt{s_{NN}} = 2.76$ TeV for Gb and Re parameters with a constant diffusion coefficient and one that is dependent on the momentum. Interpolated in the low-intermediate momentum region.

The points in Fig. (6.1) that don't have horizontal bars have a bin width of 0.5 GeV/c. Notice that the plots in Fig. (6.1) don't look quite tidy, particularly in the low transverse momentum region. For the remainder of our results, the plots will be constructed such that all bins with a width of 0.5 GeV/c will be interpolated and represented by continuous curves with bands around the curves representing the magnitude of the statistical uncertainties. All plots showing expanded views of the transverse momentum region, $0 < p_T \leq 30$ GeV/c will adopt this approach, such that they don't contain any combined bins. Note that the statistical uncertainties are small, but present in Fig. (6.1), hidden by the point-size and using continuous curves will also address this. An example of the continuous curve is shown in Fig. (6.2), which is the same as the plot shown in Fig. (6.1), with the addition of the features described in this paragraph. The motivation for this choice will become clearer when we show plots containing results of nine centrality classes, thus nine different curves.

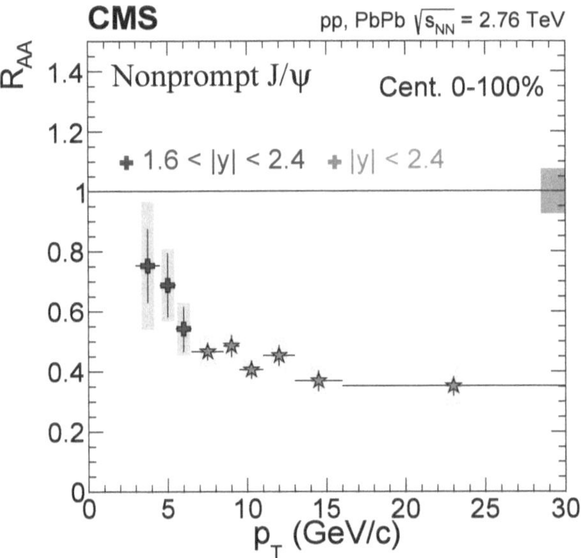

Figure 6.3: Non-prompt J/Ψ nuclear modification factor as a function of p_T at $\sqrt{s_{NN}} = 2.76$ TeV obtained in Ref. [73].

We can compare some of the qualitative features of our results in Fig. (6.2) to the CMS non-prompt J/Ψ results shown in Fig. (6.3), where the vertical bars represent statistical (systematic) uncertainties and the horizontal bars represent the bin size. Given that the CMS result covers the centrality range 0-100%, one limitation to our qualitative comparison is that the CMS nuclear modification factor result could be slightly higher than the one we've predicted since less suppression occurs in peripheral collisions. However, the CMS result covers the rapidity range $|y| < 2.4$ and the nuclear modification factor decreases with rapidity [146, 147]. These two effects will counterbalance, thus relaxing the limitations of our qualitative comparison. Another important difference is that we've given predictions for B-mesons as compared to the B decays to nonprompt J/Ψ shown in Fig. (6.3). This difference manifests by most of the J/Ψ being pushed down to low-p_T compared to B since the nonprompt J/Ψ are decay products of B-mesons and they only take a fraction of the B-meson's momentum.

Looking at Fig. (6.2), all of our predictions appear to be qualitatively consistent with the first two bins from the left of Fig. (6.3), in the transverse momentum region $3 < p_T < 5$ GeV/c. Notice that this comparison in transverse momentum has limitations as discussed in the previous paragraph, since the J/Ψ only takes a fraction of the B-meson's momentum, we expect the p_T bins in Fig. (6.3) to be shifted to the left by a few GeV's compared to the bins in Fig. (6.2). In the region $6 \leq p_T < 16$ GeV/c, the CMS data is readily described by the Gb parameters curves. However, the last bin centred at $p_T \sim 23$ GeV/c lies between the $D = const$ curves. The model based on $D(p)$ breaks down at high momentum since the longitudinal momentum fluctuations grow as $\gamma^{5/2}$, this is discussed further in Appx. (C.2).

We'll now proceed to present the nuclear modification factor results for B-mesons at $\sqrt{s_{NN}} = 5.5$ TeV. Starting with the centrality dependence of $R_{AA}(p_T)$ for each of the parameters Gb and Re with the different diffusion coefficients under consideration, Fig. (6.4) shows $R_{AA}(p_T)$ for Gb, $D(p)$ and Fig. (6.5) shows $R_{AA}(p_T)$ for Gb, $D = const$. Then Fig. (6.6) shows $R_{AA}(p_T)$ for Re, $D(p)$ and Fig. (6.7) shows $R_{AA}(p_T)$ for Re, $D = const$. Notice from looking at the plots showing the expanded view of the transverse momentum region, $0 < p_T \leq 30$ GeV/c that in all these cases employing different parameters, the nuclear modification factor peaks at a point in the interval $0 < p_T < 2$ GeV/c. The presence of this peak implies that most of the high p_T B-mesons that lose their momentum as they traverse through the QGP end up with a momentum in this region.

We can also see that less suppression is predicted for peripheral collisions, rather, there is less suppression as we move from central to peripheral collisions. This is largely influenced by the initial geometry of the overlap region, in that, if the overlap region is smaller (i.e. in peripheral collisions) then less QGP is produced and the heavy quarks spend very little time in the QGP medium and lose less energy compared to central collisions. This is also explained by the results shown in Fig. (4.2 - 4.5) where a comparison is made on the time spent by a heavy quark in the QGP medium if it is produced at a certain point (with a certain momentum) in the medium at different centralities.

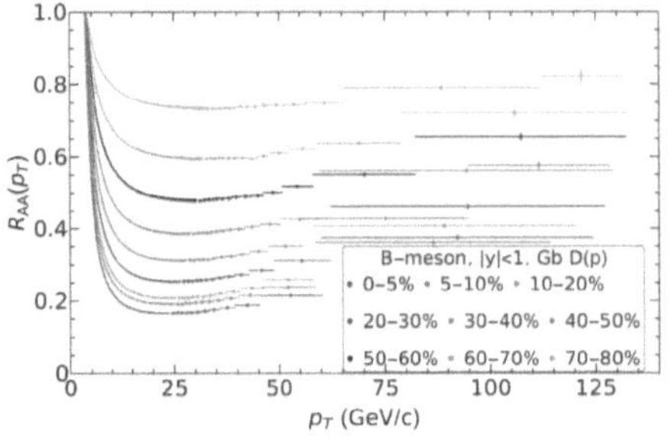
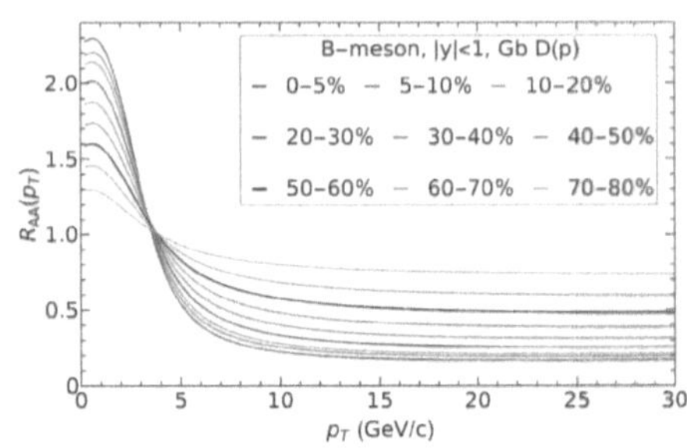

(a) *Gb*, $D(p)$ B-meson nuclear modification factor at $\sqrt{s_{NN}} = 5.5$ TeV for centrality classes 0-5% up to 70-80%.

(b) Expanded view of the transverse momentum region, $0 < p_T \leq 30$ GeV/c of Fig. (a), on the left, including the region $R_{AA}(p_T) > 1$.

Figure 6.4: B-meson nuclear modification factor at $\sqrt{s_{NN}} = 5.5$ TeV for *Gb* parameters with a diffusion coefficient that is dependent on momentum, $D(p)$.

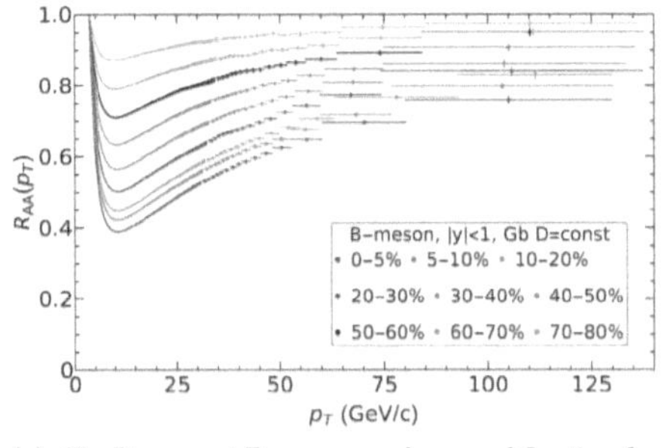
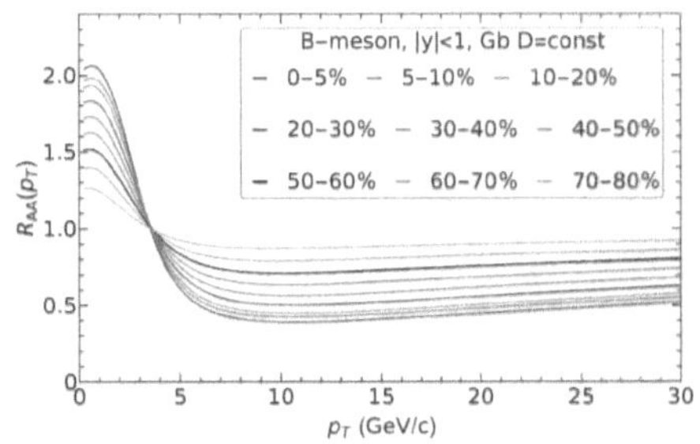

(a) *Gb*, $D = const$ B-meson nuclear modification factor at $\sqrt{s_{NN}} = 5.5$ TeV for centrality classes 0-5% up to 70-80%.

(b) Expanded view of the transverse momentum region, $0 < p_T \leq 30$ GeV/c of Fig. (a), on the left, including the region $R_{AA}(p_T) > 1$.

Figure 6.5: B-meson nuclear modification factor at $\sqrt{s_{NN}} = 5.5$ TeV for *Gb* parameters with a diffusion coefficient that is not dependent on momentum, $D = const$.

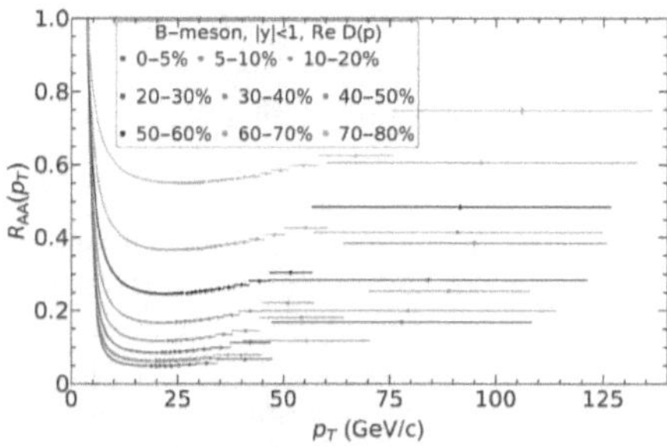
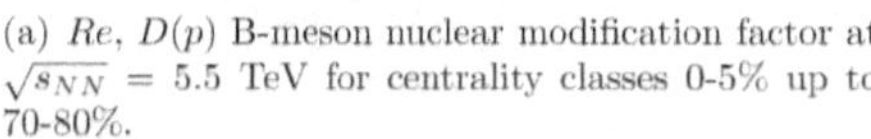
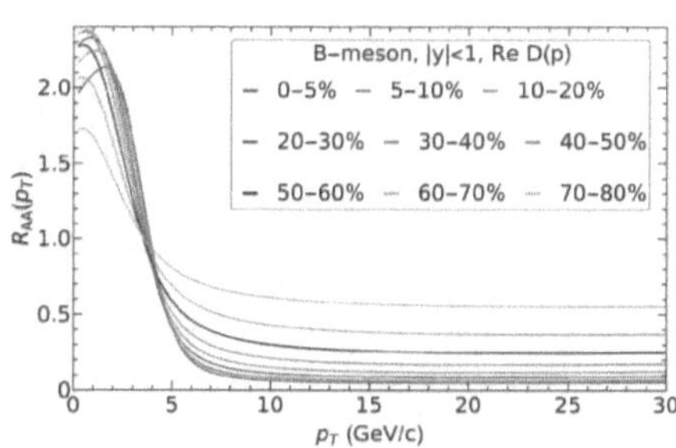

(a) *Re*, $D(p)$ B-meson nuclear modification factor at $\sqrt{s_{NN}} = 5.5$ TeV for centrality classes 0-5% up to 70-80%.

(b) Expanded view of the transverse momentum region, $0 < p_T \leq 30$ GeV/c of Fig. (a), on the left, including the region $R_{AA}(p_T) > 1$.

Figure 6.6: B-meson nuclear modification factor at $\sqrt{s_{NN}} = 5.5$ TeV for *Re* parameters with a diffusion coefficient that is dependent on momentum, $D(p)$.

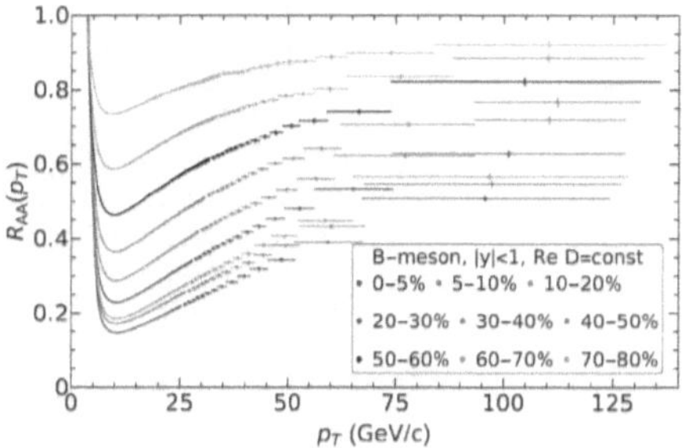
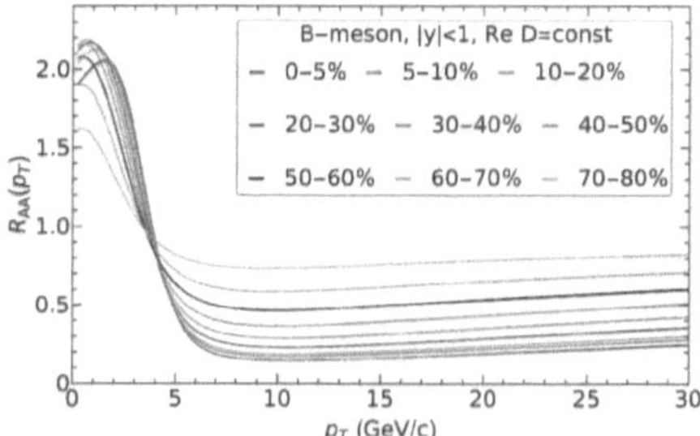

(a) *Re*, $D = const$ B-meson nuclear modification factor at $\sqrt{s_{NN}} = 5.5$ TeV for centrality classes 0-5% up to 70-80%.

(b) Expanded view of the transverse momentum region, $0 < p_T \leq 30$ GeV/c of Fig. (a), on the left, including the region $R_{AA}(p_T) > 1$.

Figure 6.7: B-meson nuclear modification factor at $\sqrt{s_{NN}} = 5.5$ TeV for *Re* parameters with a diffusion coefficient that is not dependent on momentum, $D = const$.

We now look at the nuclear modification factor per centrality class to see how $R_{AA}(p_T)$ varies for the different parameters we've employed. These results are shown in Fig. (6.8 -6.16) for the different centralities. Notice first that each of the curves cross at some point in the p_T domain covered, so there is a point where the predictions of the models agree in the low-intermediate momentum region. However, as we move out to the high transverse momentum region, the over-suppression of the models where the diffusion coefficient is dependent on momentum,

$D(p)$ becomes more pronounced. This over-suppression is because heavy quark models where the diffusion coefficient is dependent on their velocity break down at high momenta since the momentum fluctuations grow quickly with momentum in the longitudinal direction, such that they contribute immensely to the heavy quark energy loss.

The drag coefficient, μ in Eq. (4.2) has the largest contribution to the energy loss, in the Gb prescription, the 't Hooft coupling is smaller by ≈ 2 and T is lower, so the drag for Gb parameters is smaller compared to Re parameters and results in less suppression. This difference in μ for Gb and Re parameters is clearly reflected in our results shown in Fig. (6.8 - 6.16) as the Gb curves show a higher $R_{AA}(p_T)$ compared to Re curves for the same diffusion coefficient. We show the comparison of the drag coefficients for Gb and Re parameters in Eq. (6.1),

$$\frac{\pi\sqrt{5.5}T^2}{2M} < \frac{\pi\sqrt{11.3}(3^{1/4}T)^2}{2M}, \quad \implies \mu_{Gb} < \mu_{Re} \tag{6.1}$$

where the LHS corresponds to Gb and the RHS to Re parameters respectively and it's clear that $\mu_{Gb} < \mu_{Re}$. Taking another look at the results presented in Fig. (6.8 - 6.16), one can also see that the different curves for different parameters and in different centrality classes all cross $R_{AA}(p_T) = 1$ at similar points in $3 < p_T < 5$ GeV/c. This similarity in the p_T value where $R_{AA}(p_T) = 1$ is a result of the power law shape of the b-quark production spectrum which falls off as $1/(ap_T^4 + \text{lower order terms})$, where a is a constant.

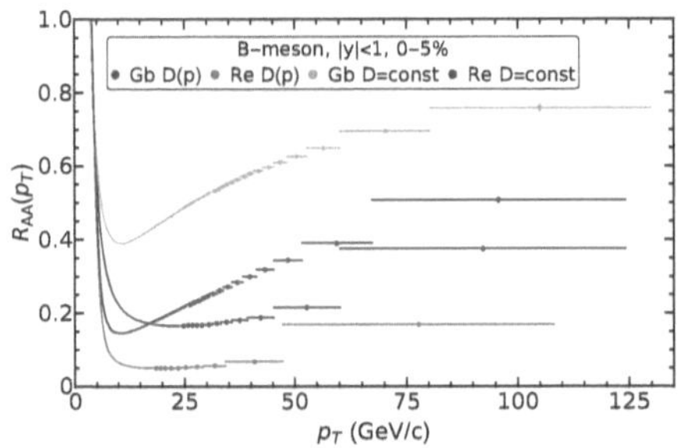
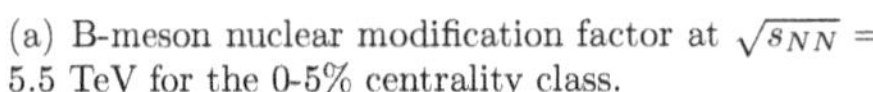
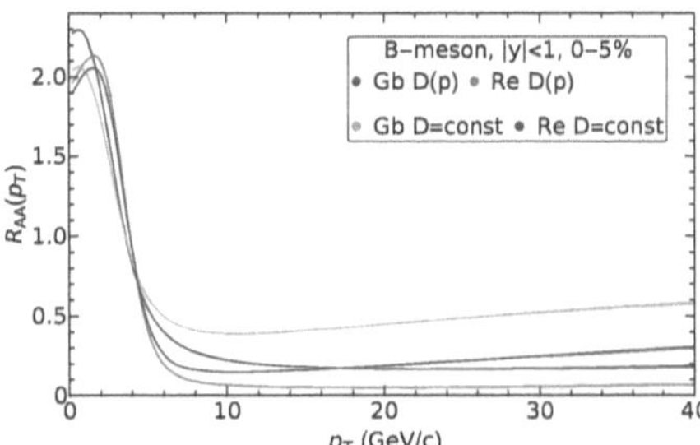

(a) B-meson nuclear modification factor at $\sqrt{s_{NN}} =$ 5.5 TeV for the 0-5% centrality class.

(b) Expanded view of the transverse momentum region, $0 < p_T \leq 30$ GeV/c of Fig. (a), on the left, including the region $R_{AA}(p_T) > 1$.

Figure 6.8: B-meson nuclear modification factor at $\sqrt{s_{NN}} = 5.5$ TeV for Gb and Re parameters with a constant diffusion coefficient and one that is dependent on the momentum at 0-5% centrality.

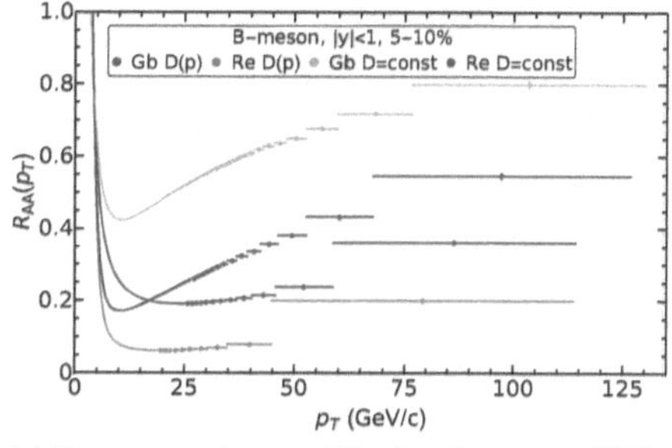

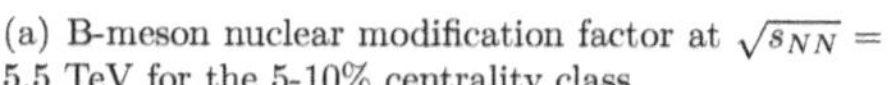

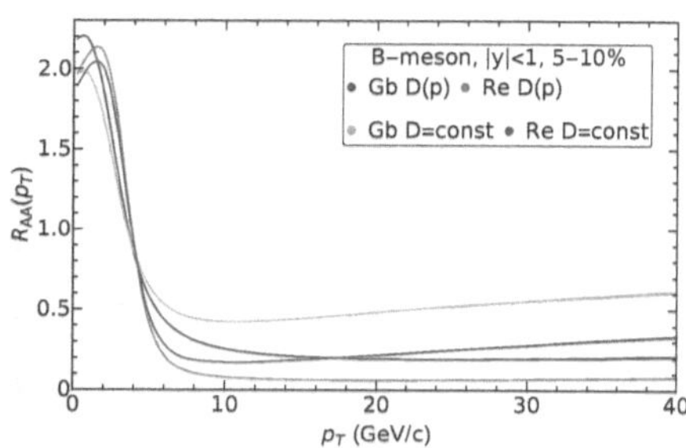

(a) B-meson nuclear modification factor at $\sqrt{s_{NN}} =$ 5.5 TeV for the 5-10% centrality class.

(b) Expanded view of the transverse momentum region, $0 < p_T \leq 30$ GeV/c of Fig. (a), on the left, including the region $R_{AA}(p_T) > 1$.

Figure 6.9: B-meson nuclear modification factor at $\sqrt{s_{NN}} = 5.5$ TeV for Gb and Re parameters with a constant diffusion coefficient and one that is dependent on the momentum at 5-10% centrality.

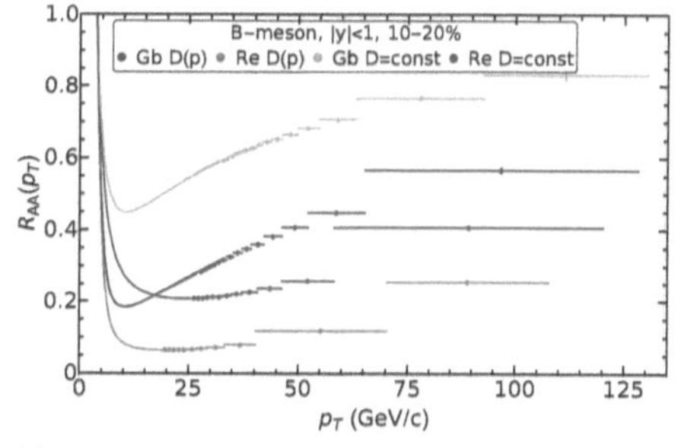

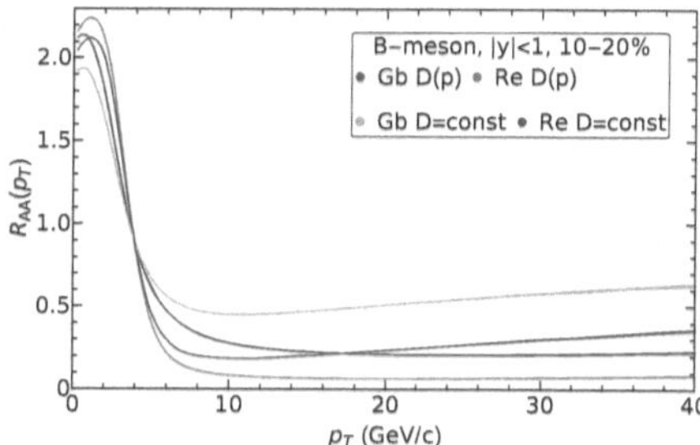

(a) B-meson nuclear modification factor at $\sqrt{s_{NN}} =$ 5.5 TeV for the 10-20% centrality class.

(b) Expanded view of the transverse momentum region, $0 < p_T \leq 30$ GeV/c of Fig. (a), on the left, including the region $R_{AA}(p_T) > 1$.

Figure 6.10: B-meson nuclear modification factor at $\sqrt{s_{NN}} = 5.5$ TeV for Gb and Re parameters with a constant diffusion coefficient and one that is dependent on the momentum at 10-20% centrality.

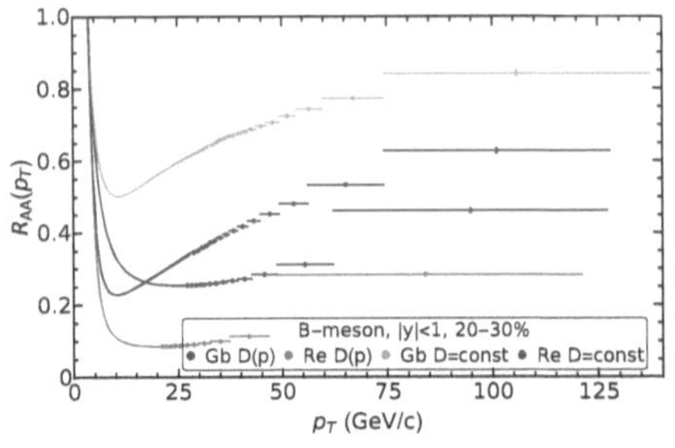
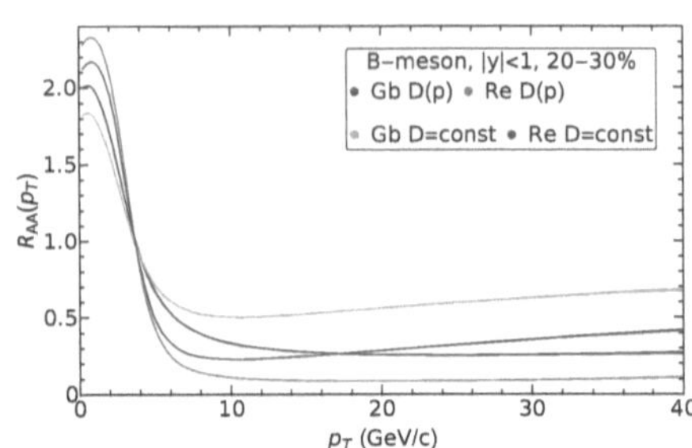

(a) B-meson nuclear modification factor at $\sqrt{s_{NN}} =$ 5.5 TeV for the 20-30% centrality class.

(b) Expanded view of the transverse momentum region, $0 < p_T \leq 30$ GeV/c of Fig. (a), on the left, including the region $R_{AA}(p_T) > 1$.

Figure 6.11: B-meson nuclear modification factor at $\sqrt{s_{NN}} = 5.5$ TeV for Gb and Re parameters with a constant diffusion coefficient and one that is dependent on the momentum at 20-30% centrality.

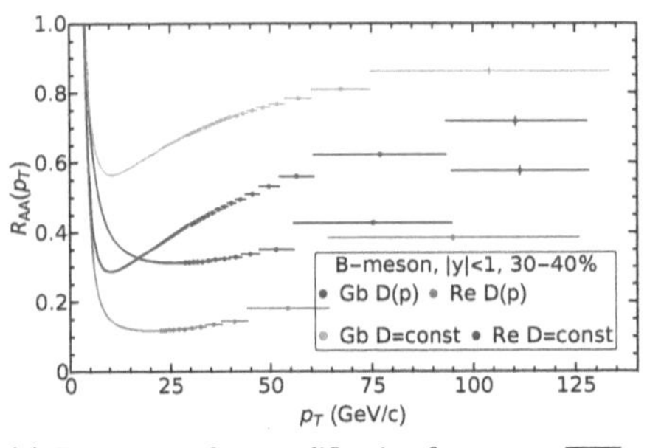
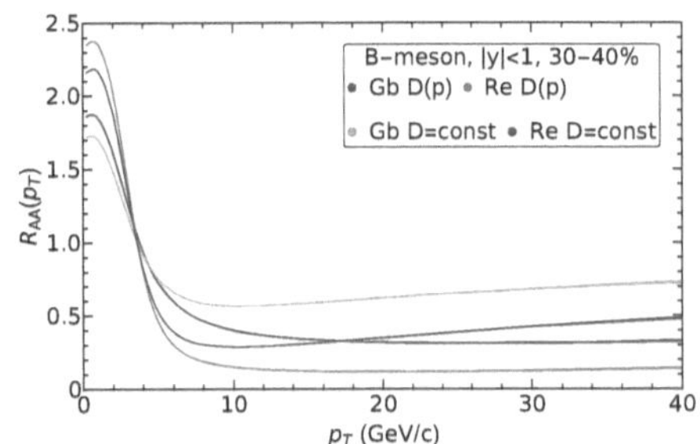

(a) B-meson nuclear modification factor at $\sqrt{s_{NN}} =$ 5.5 TeV for the 30-40% centrality class.

(b) Expanded view of the transverse momentum region, $0 < p_T \leq 30$ GeV/c of Fig. (a), on the left, including the region $R_{AA}(p_T) > 1$.

Figure 6.12: B-meson nuclear modification factor at $\sqrt{s_{NN}} = 5.5$ TeV for Gb and Re parameters with a constant diffusion coefficient and one that is dependent on the momentum at 30-40% centrality.

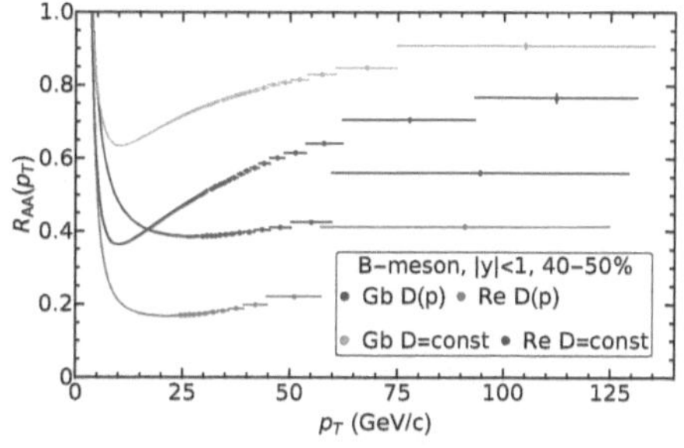
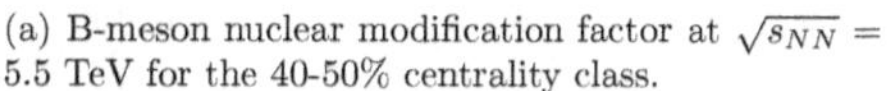
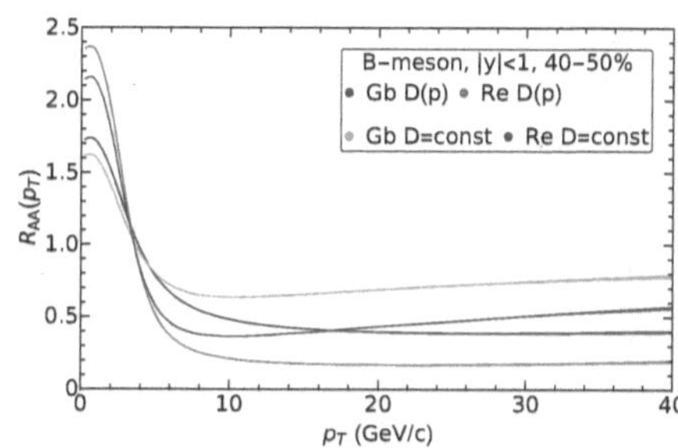

(a) B-meson nuclear modification factor at $\sqrt{s_{NN}} =$ 5.5 TeV for the 40-50% centrality class.

(b) Expanded view of the transverse momentum region, $0 < p_T \leq 30$ GeV/c of Fig. (a), on the left, including the region $R_{AA}(p_T) > 1$.

Figure 6.13: B-meson nuclear modification factor at $\sqrt{s_{NN}} = 5.5$ TeV for Gb and Re parameters with a constant diffusion coefficient and one that is dependent on the momentum at 40-50% centrality.

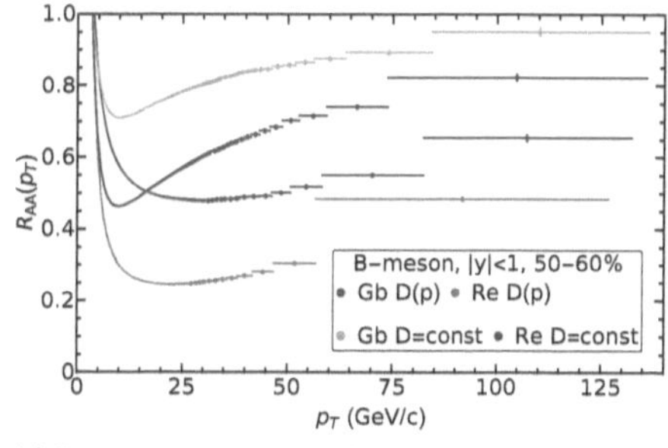
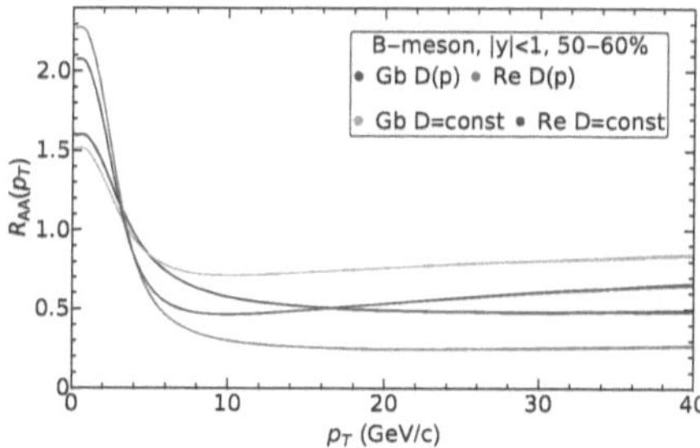

(a) B-meson nuclear modification factor at $\sqrt{s_{NN}} =$ 5.5 TeV for the 50-60% centrality class.

(b) Expanded view of the transverse momentum region, $0 < p_T \leq 30$ GeV/c of Fig. (a), on the left, including the region $R_{AA}(p_T) > 1$.

Figure 6.14: B-meson nuclear modification factor at $\sqrt{s_{NN}} = 5.5$ TeV for Gb and Re parameters with a constant diffusion coefficient and one that is dependent on the momentum at 50-60% centrality.

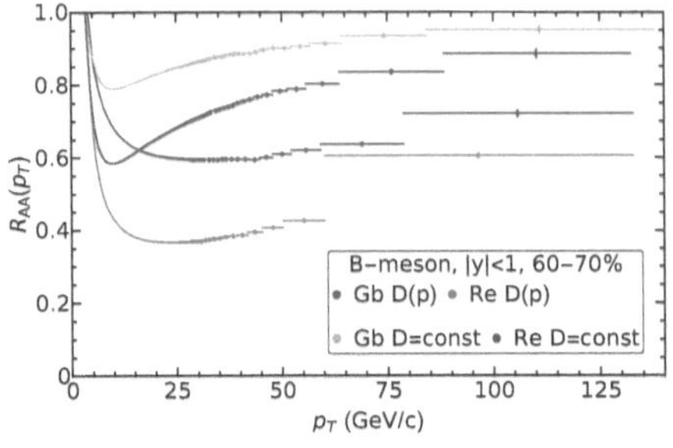 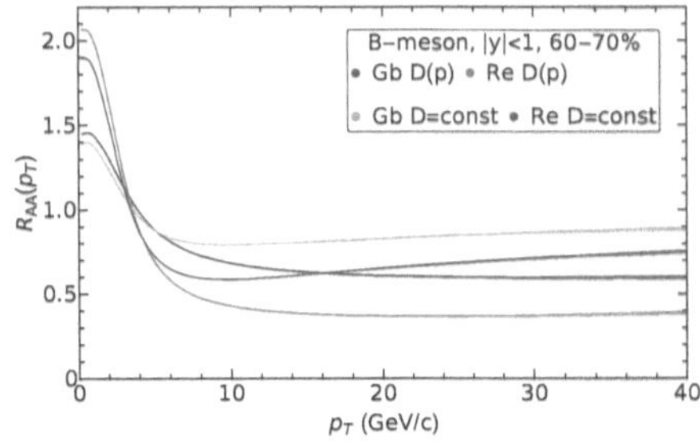

(a) B-meson nuclear modification factor at $\sqrt{s_{NN}} =$ 5.5 TeV for the 60-70% centrality class.

(b) Expanded view of the transverse momentum region, $0 < p_T \leq 30$ GeV/c of Fig. (a), on the left, including the region $R_{AA}(p_T) > 1$.

Figure 6.15: B-meson nuclear modification factor at $\sqrt{s_{NN}} = 5.5$ TeV for Gb and Re parameters with a constant diffusion coefficient and one that is dependent on the momentum at 60-70% centrality.

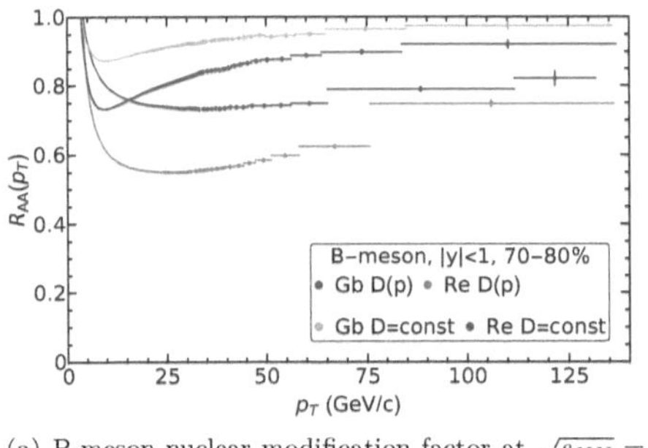 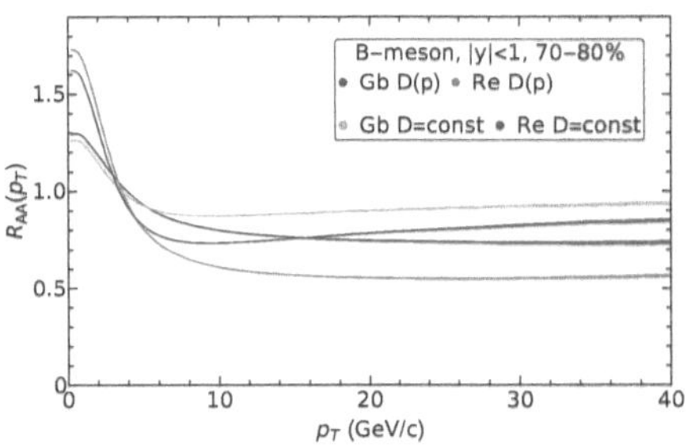

(a) B-meson nuclear modification factor at $\sqrt{s_{NN}} =$ 5.5 TeV for the 70-80% centrality class.

(b) Expanded view of the transverse momentum region, $0 < p_T \leq 30$ GeV/c of Fig. (a), on the left, including the region $R_{AA}(p_T) > 1$.

Figure 6.16: B-meson nuclear modification factor at $\sqrt{s_{NN}} = 5.5$ TeV for Gb and Re parameters with a constant diffusion coefficient and one that is dependent on the momentum at 70-80% centrality.

6.2 $v_2(p_T)$

In Fig. (6.17) we show predictions for the $v_2(p_T)$ for B-mesons at $\sqrt{s_{NN}} = 2.76$ TeV at $0-10\%$ centrality in the rapidity range $|y| < 0.5$. The inset plot shows an expanded view of the transverse momentum region, $0 < p_T \leq 30$ GeV/c. The vertical bars and error bands represent statistical uncertainties computed using the method described in Chap. (5.2.1) while the horizontal bars indicate the bin width. We can make a qualitative comparison of these predictions to the CMS nonprompt J/Ψ experimental results shown in Fig. (6.18). This qualitative comparison has some limitations, we are comparing B-meson predictions to nonprompt J/Ψ data and the nonprompt J/Ψ bins are shifted to lower p_T since the J/Ψ only takes a fraction of the B-meson's momentum. We expect the CMS $v_2(p_T)$ to be slightly higher than the $v_2(p_T)$ we've predicted since $v_2(p_T)$ increases in semi-central collisions [148]. The CMS data also covers a wider rapidity range, however, we don't expect to see this limitation manifesting since v_2 doesn't have a strong rapidity dependence [148, 149].

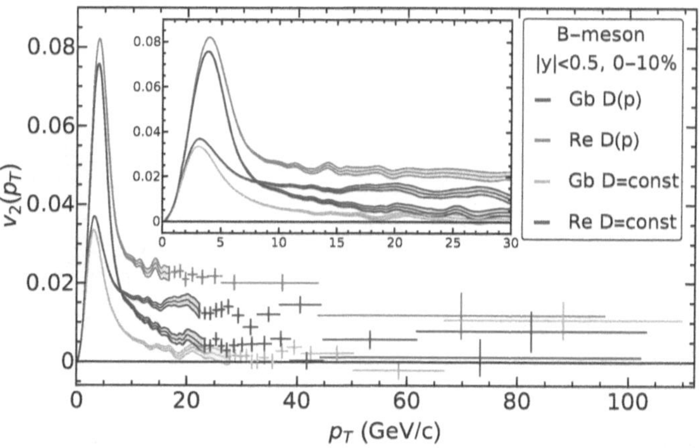

Figure 6.17: $v_2(p_T)$ for B-mesons at $\sqrt{s_{NN}} = 2.76$ TeV with diffusion coefficients, $D = const$ as well as $D(p)$, for both Gb and Re parameters at 0-10% centrality.

Looking at Fig. (6.18), we see that the left bin on $3 < p_T < 6.5$ GeV/c in the rapidity range $1.6 < |y| < 2.4$ is in the region between the $Re\ D(p)$, $Re\ D = const$ and $Gb\ D(p)$ curves. The right bin on $6.5 < p_T < 30$ GeV/c in the rapidity range $|y| < 2.4$ lies in the region between the $Gb\ D(p)$ and $Re\ D(p)$ curves. The experimental v_2 as seen from the two bins in Fig. (6.18) is slightly higher since it covers a larger centrality range including mid-central collisions where the

v_2 is generally higher due to the high initial spatial anisotropy. Therefore, the $D(p)$ parameters $v_2(p_T)$ predictions are qualitatively consistent with this experimental data, while the $D = const$ predictions are not consistent since the $v_2(p_T)$ computed using these parameters is very low at intermediate-large transverse momentum.

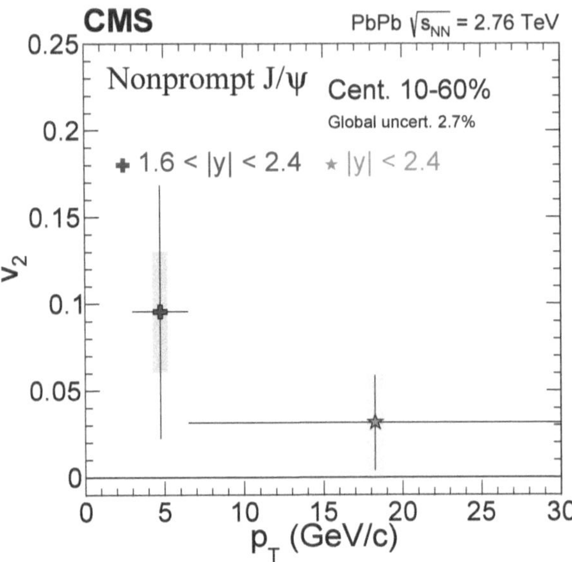

Figure 6.18: Nonprompt J/Ψ $v_2(p_T)$ as a function of p_T at $\sqrt{s_{NN}} = 2.76$ TeV measured at CMS, obtained in Ref. [73].

Our $v_2(p_T)$ result, shown in Fig. (6.17), increases sharply as the $R_{AA}(p_T)$ shown in Fig. (6.2) decreases sharply and $v_2(p_T)$ peaks around $p_T \sim 3$ GeV/c for Gb parameters and $p_T \sim 4$ GeV/c for Re parameters. The $v_2(p_T)$ peak we observe in our predictions at intermediate transverse momentum, $3 < p_T < 4$ GeV/c, as well as the drop in $v_2(p_T)$ that we observe at high p_T is consistent with results obtained in [150]. A possible reason for the $v_2(p_T)$ peak, based on hydrodynamics (note that we've used hydrodynamic backgrounds) is discussed in Ref. [151] as follows: at freeze-out, the hydrodynamic flow is on average stronger in the reaction plane than the plane perpendicular to it, at high collision energies, the flow velocity attains its largest value on the freeze-out surface in the out-of-plane direction. Given that the highest transverse momentum hadrons are emitted from fluid cells with the highest flow velocities, as a result, the $v_2(p_T)$ is reduced at high transverse momentum and turns negative in some cases, as can be seen for the *Gb D = const* $50 < p_T < 67$ GeV/c bin in Fig. (6.17).

Some qualitative features in our $v_2(p_T)$ results can be explained by hydrodynamics since we are propagating the heavy quarks through a medium based on a hydrodynamic model. We can also explain these features using our energy-loss model, for example, by looking at the coupling. There's a clear anti-correlation between the $v_2(p_T)$ and $R_{AA}(p_T)$, the lowest $v_2(p_T)$ (i.e. *Gb D = const*) corresponds to the highest $R_{AA}(p_T)$ and this can be seen on Fig. (6.2). Since the production angle of the heavy quarks is uniformly distributed, in the event that the heavy quarks don't couple to the medium (i.e. coupling is zero), then there would be no energy loss and the nuclear modification factor would be one, consequently there would be no $v_2(p_T)$, i.e. $v_2(p_T) = 0$. If we increase the coupling, there will be more energy loss and the nuclear modification factor will decrease, as a result the $v_2(p_T)$ increases.

We'll now look at $v_2(p_T)$ results at $\sqrt{s_{NN}} = 5.5$ TeV for the same centrality classes in which we studied the $R_{AA}(p_T)$ in Sec. (6.1). In Fig. (6.19 - 6.26) we show the B-meson $v_2(p_T)$ predictions for all the parameters under consideration, covering central, semi-central and peripheral collisions. The $v_2(p_T)$ is low for central collisions, dropping to negative values in some instances such as in Fig. (6.21) and increases as we move up in centrality to mid-central collisions. This change in $v_2(p_T)$ with centrality is caused by the initial geometry of the collision as discussed in Chap. (3). The shape of the overlap region is more symmetrical in central collisions as shown in Fig. (3.6), so there's very little spatial anisotropy, resulting in the low momentum anisotropy. On the other hand, the overlap region is almond shaped for mid-central to peripheral collisions as shown in Fig. (3.11, 3.12) and this large spatial anisotropy converts to a large momentum anisotropy and consequently large v_2.

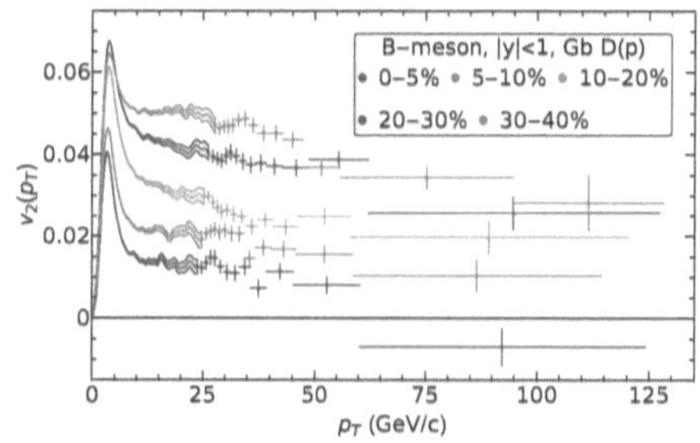

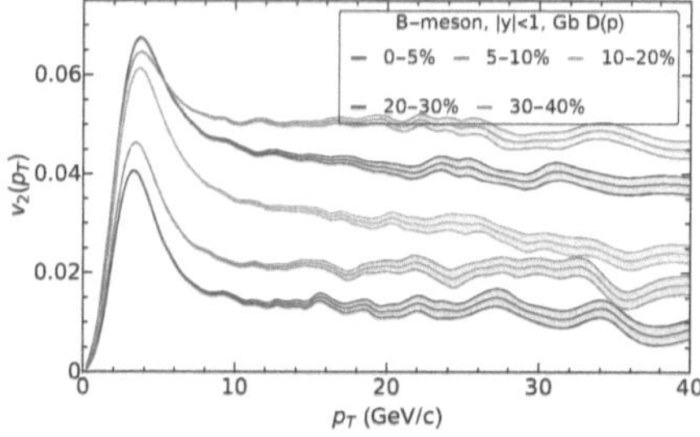

(a) *Gb*, *D(p)* B-meson $v_2(p_T)$ at $\sqrt{s_{NN}} = 5.5$ TeV for centrality classes 0-5% up to 30-40%.

(b) Expanded view of the transverse momentum region, $0 < p_T \leq 30$ GeV/c of Fig. (a), on the left.

Figure 6.19: $v_2(p_T)$ for B-mesons at $\sqrt{s_{NN}} = 5.5$ TeV for central and mid-central collisions with a diffusion coefficient that is dependent on momentum, $D(p)$, for *Gb* parameters.

The largest $v_2(p_T)$ occurs in mid-central collisions, i.e. the $30 - 40\%$ centrality class where the impact parameter is approximately the size of the nuclear radius and the spatial asymmetry of the collision is at its largest. This relatively large $v_2(p_T)$ at $30 - 40\%$ centrality can be seen in all the figures showing $v_2(p_T)$ for central to semi-central collisions for the various parameters, such as Fig. (6.19). The $v_2(p_T)$ decreases for peripheral collisions as a result of a lack of collectivity in the hydrodynamic medium [152], hence the peak at $3 < p_T < 5$ GeV/c becomes less pronounced and eventually disappears for some of the parameters such as Fig. (6.20).

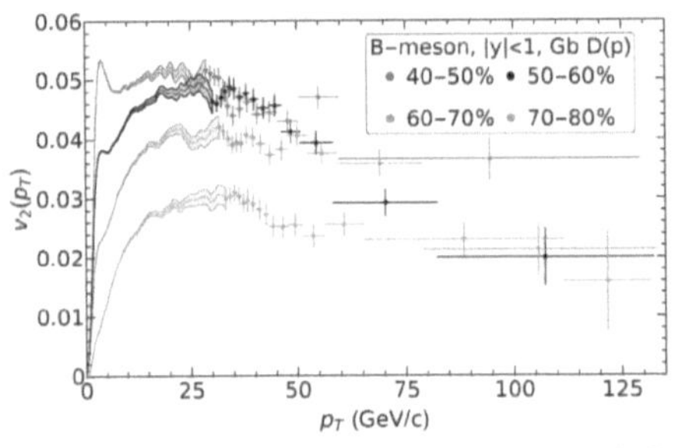

(a) *Gb*, $D(p)$ B-meson $v_2(p_T)$ at $\sqrt{s_{NN}} = 5.5$ TeV for centrality classes 40-50% up to 70-80%.

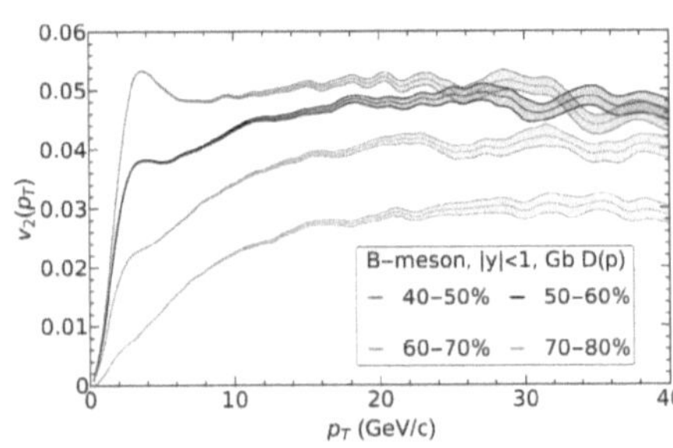

(b) Expanded view of the transverse momentum region, $0 < p_T \leq 30$ GeV/c of Fig. (a), on the left.

Figure 6.20: $v_2(p_T)$ for B-mesons at $\sqrt{s_{NN}} = 5.5$ TeV for peripheral collisions with a diffusion coefficient that is dependent on momentum, $D(p)$, for *Gb* parameters.

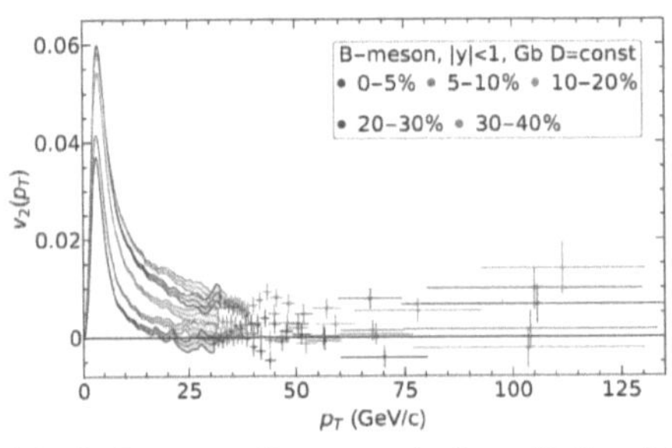

(a) *Gb*, $D = const$ B-meson $v_2(p_T)$ at $\sqrt{s_{NN}} = 5.5$ TeV for centrality classes 0-5% up to 30-40%.

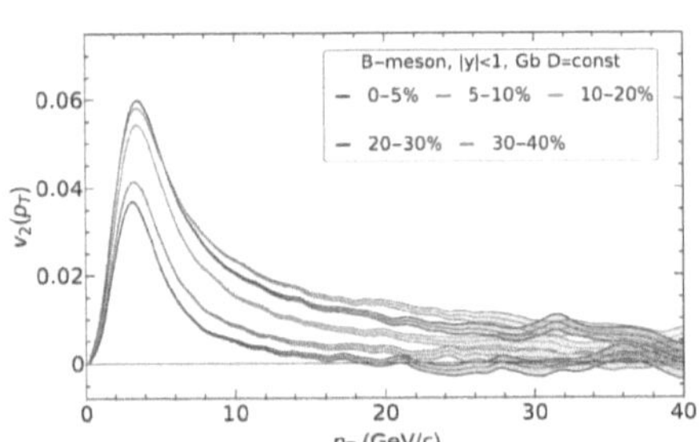

(b) Expanded view of the transverse momentum region, $0 < p_T \leq 30$ GeV/c of Fig. (a), on the left.

Figure 6.21: $v_2(p_T)$ for B-mesons at $\sqrt{s_{NN}} = 5.5$ TeV for central and mid-central collisions with a diffusion coefficient that is not dependent on momentum, $D = const$, for *Gb* parameters.

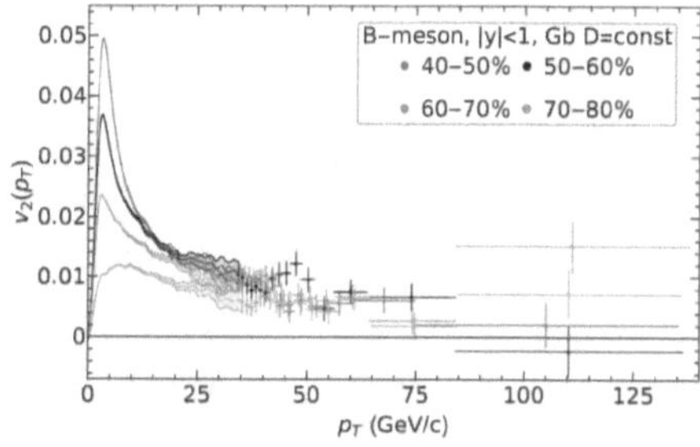
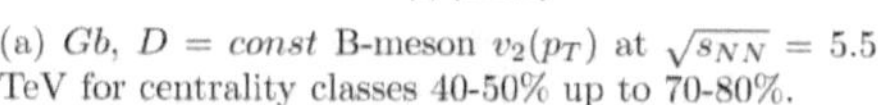
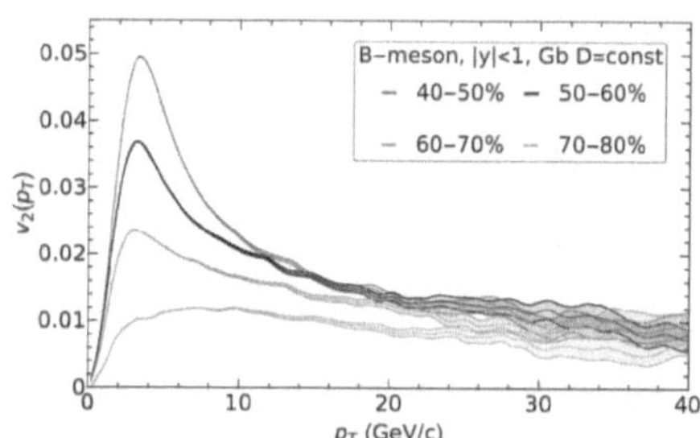

(a) *Gb*, $D = const$ B-meson $v_2(p_T)$ at $\sqrt{s_{NN}} = 5.5$ TeV for centrality classes 40-50% up to 70-80%.

(b) Expanded view of the transverse momentum region, $0 < p_T \leq 30$ GeV/c of Fig. (a), on the left.

Figure 6.22: $v_2(p_T)$ for B-mesons at $\sqrt{s_{NN}} = 5.5$ TeV for peripheral collisions with a diffusion coefficient that is not dependent on momentum, $D = const$, for *Gb* parameters.

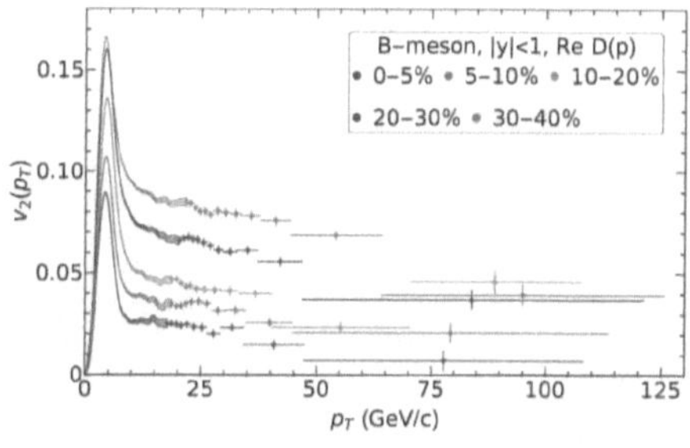
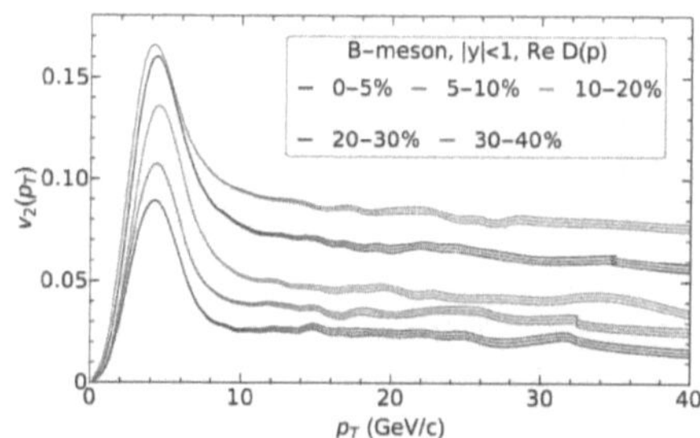

(a) *Re*, $D(p)$ B-meson $v_2(p_T)$ at $\sqrt{s_{NN}} = 5.5$ TeV for centrality classes 0-5% up to 30-40%.

(b) Expanded view of the transverse momentum region, $0 < p_T \leq 30$ GeV/c of Fig. (a), on the left.

Figure 6.23: $v_2(p_T)$ for B-mesons at $\sqrt{s_{NN}} = 5.5$ TeV for central and mid-central collisions with a diffusion coefficient that is dependent on momentum, $D(p)$, for *Re* parameters.

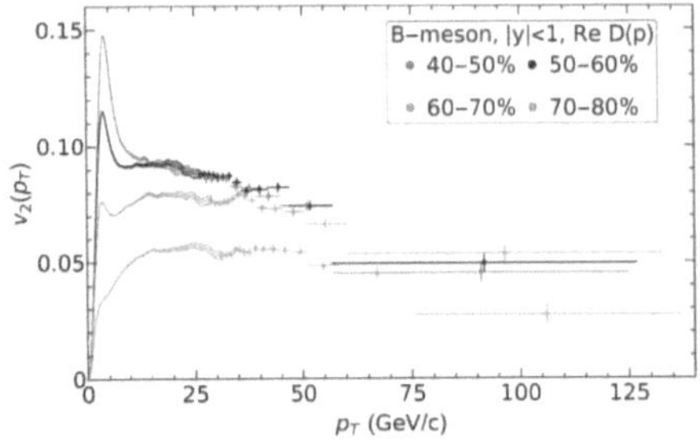
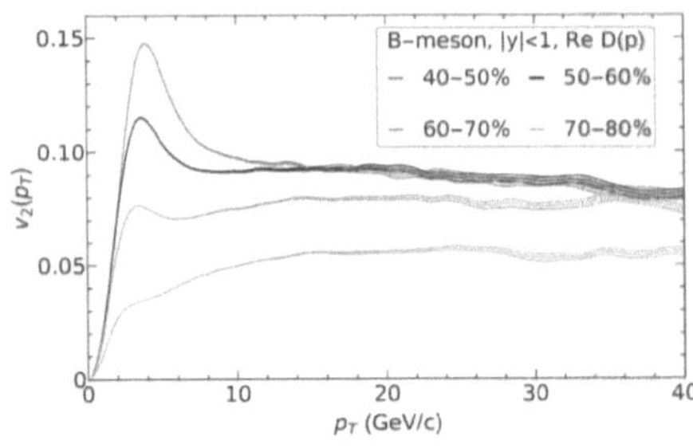

(a) Re, $D(p)$ B-meson $v_2(p_T)$ at $\sqrt{s_{NN}} = 5.5$ TeV for centrality classes 40-50% up to 70-80%.

(b) Expanded view of the transverse momentum region, $0 < p_T \leq 30$ GeV/c of Fig. (a), on the left.

Figure 6.24: $v_2(p_T)$ for B-mesons at $\sqrt{s_{NN}} = 5.5$ TeV for peripheral collisions with a diffusion coefficient that is dependent on momentum, $D(p)$, for Re parameters.

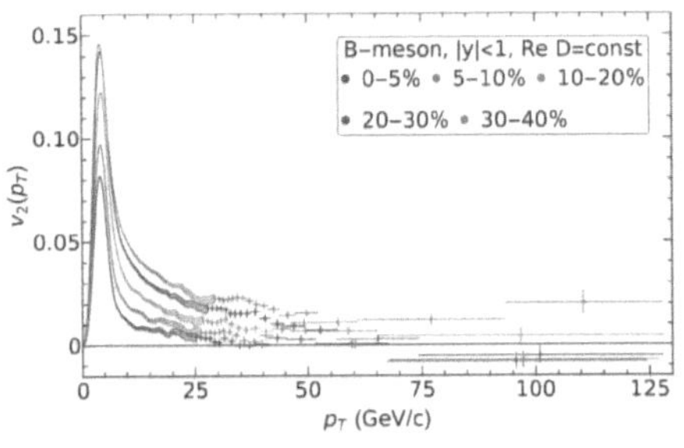
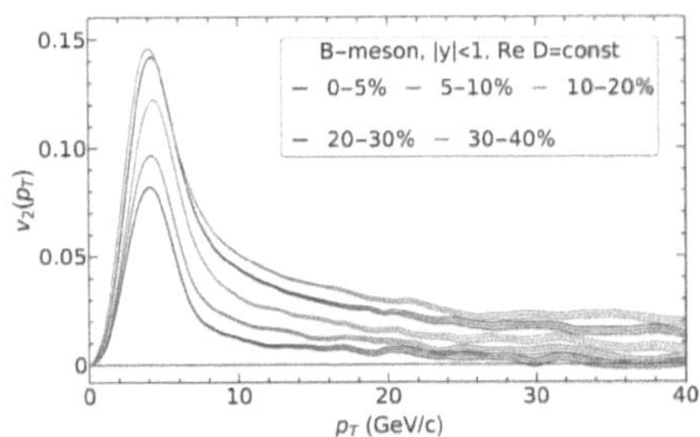

(a) Re. $D = const$ B-meson $v_2(p_T)$ at $\sqrt{s_{NN}} = 5.5$ TeV for centrality classes 0-5% up to 30-40%.

(b) Expanded view of the transverse momentum region, $0 < p_T \leq 30$ GeV/c of Fig. (a), on the left.

Figure 6.25: $v_2(p_T)$ for B-mesons at $\sqrt{s_{NN}} = 5.5$ TeV for central and mid-central collisions with a diffusion coefficient that is not dependent on momentum, $D = const$, for Re parameters.

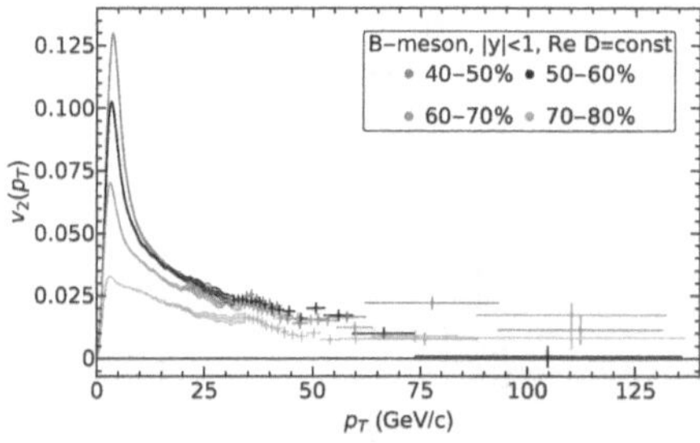
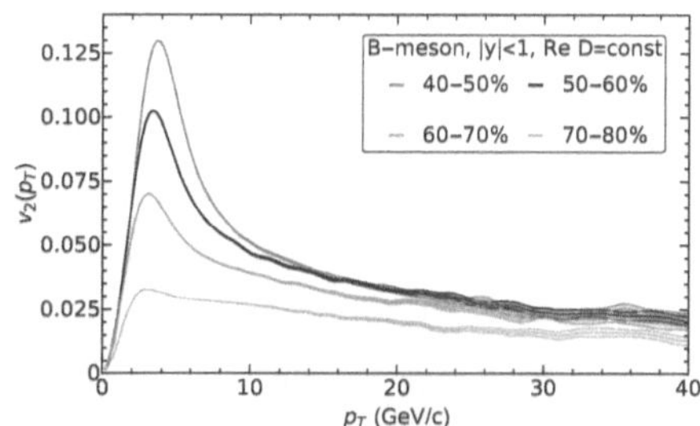

(a) *Re, D = const* B-meson $v_2(p_T)$ at $\sqrt{s_{NN}} = 5.5$ TeV for centrality classes 40-50% up to 70-80%.

(b) Expanded view of the transverse momentum region, $0 < p_T \leq 30$ GeV/c of Fig. (a), on the left.

Figure 6.26: $v_2(p_T)$ for B-mesons at $\sqrt{s_{NN}} = 5.5$ TeV for peripheral collisions with a diffusion coefficient that is not dependent on momentum, $D = const$, for *Re* parameters.

Having discussed the centrality dependence of our $v_2(p_T)$ predictions, we now compare the $v_2(p_T)$ predictions for each set of parameters per centrality class. These results are shown in Fig. (6.27 - 6.33). Notice the anti-correlation of these $v_2(p_T)$ predictions to the $R_{AA}(p_T)$ results discussed in Sec. (6.1). We obtain the largest $v_2(p_T)$ for *Re, D(p)* parameters, which correspond to the lowest nuclear modification factor and the lowest $v_2(p_T)$ occurs for *Gb, D = const* parameters corresponding to the largest nuclear modification factor.

We see that our $v_2(p_T)$ predictions at $\sqrt{s_{NN}} = 5.5$ TeV as shown in Fig. (6.27) and Fig. (6.28 a) are slightly higher compared to predictions made at $\sqrt{s_{NN}} = 2.76$ TeV shown in Fig. (6.17) for similar rapidities. This increase in $v_2(p_T)$ at $\sqrt{s_{NN}} = 5.5$ TeV is expected as data [16] shows that $v_2(p_T)$ increases with beam energy. The reason for this increase in $v_2(p_T)$ is that since $v_2(p_T)$ is generated early in the QGP expansion [153], collisions at higher energies produce a hotter system which has a partonic phase that lives for a longer period, thus contributing more to the $v_2(p_T)$. Another possible reason for the increase in $v_2(p_T)$ at higher energies is the increase in the average transverse momentum of the produced particles. This increase in the contribution of the partonic phase to the $v_2(p_T)$ in a hotter QGP medium is crucial in improving our understanding of the partonic-dominated phase of the QGP medium.

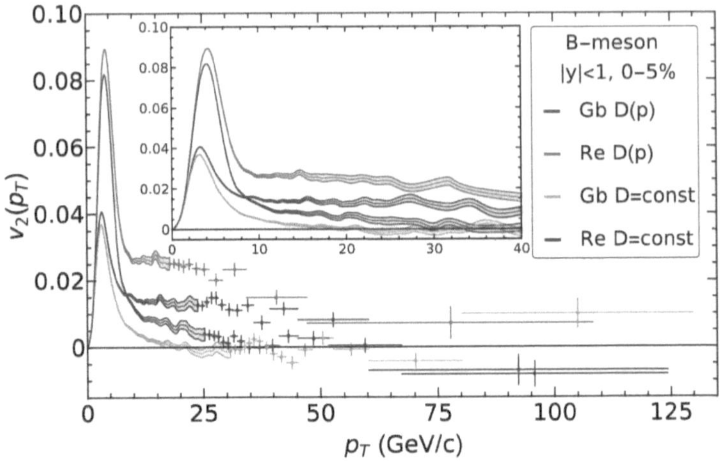

Figure 6.27: $v_2(p_T)$ for B-mesons at $\sqrt{s_{NN}} = 5.5$ TeV with diffusion coefficients, $D = const$ as well as $D(p)$, for both *Gb* and *Re* parameters at 0-5% centrality. The inset plot shows an expanded view of the low-intermediate momentum region.

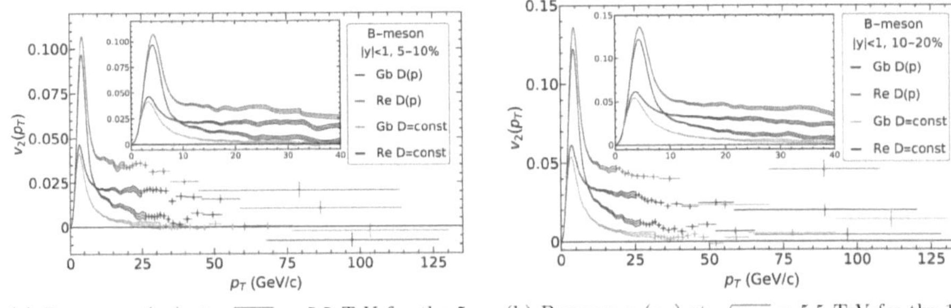

(a) B-meson $v_2(p_T)$ at $\sqrt{s_{NN}} = 5.5$ TeV for the 5-10% centrality class.

(b) B-meson $v_2(p_T)$ at $\sqrt{s_{NN}} = 5.5$ TeV for the 10-20% centrality class.

Figure 6.28: $v_2(p_T)$ for B-mesons at $\sqrt{s_{NN}} = 5.5$ TeV for both *Gb* and *Re* parameters with diffusion coefficients, $D = const$ as well as $D(p)$ for central collisions. The inset plot shows an expanded view of the low-intermediate momentum region.

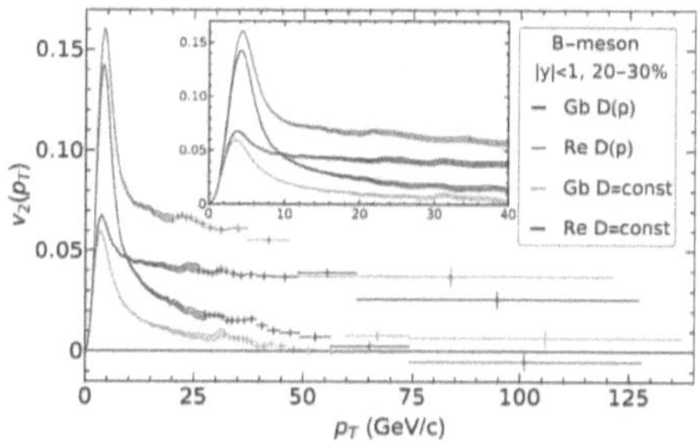

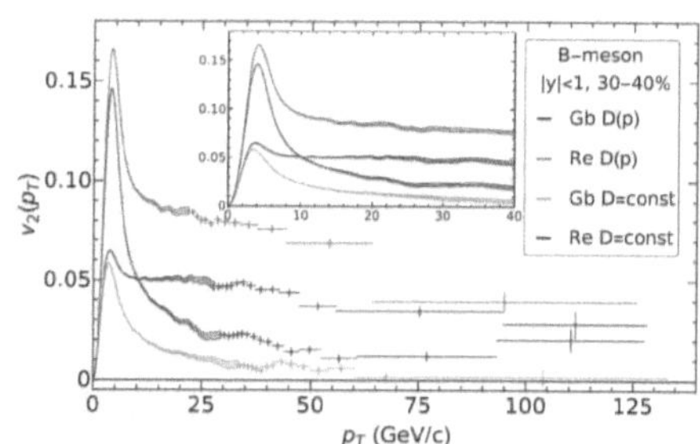

(a) B-meson $v_2(p_T)$ at $\sqrt{s_{NN}} = 5.5$ TeV for the 20-30% centrality class.

(b) B-meson $v_2(p_T)$ at $\sqrt{s_{NN}} = 5.5$ TeV for the 30-40% centrality class.

Figure 6.29: $v_2(p_T)$ for B-mesons at $\sqrt{s_{NN}} = 5.5$ TeV for both *Gb* and *Re* parameters with diffusion coefficients, $D = const$ as well as $D(p)$ for semi-central collisions. The inset plot shows an expanded view of the low-intermediate momentum region.

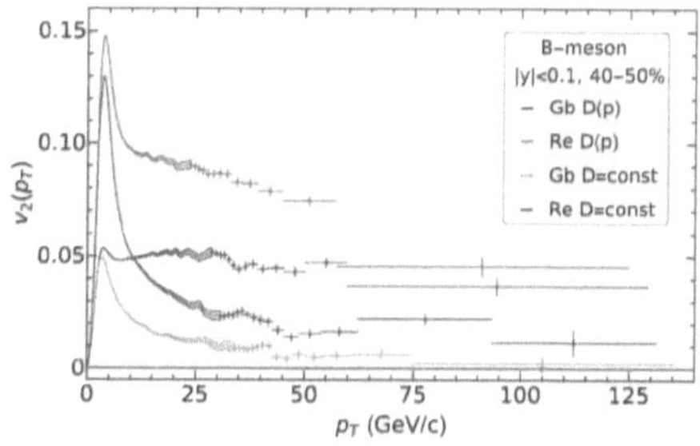

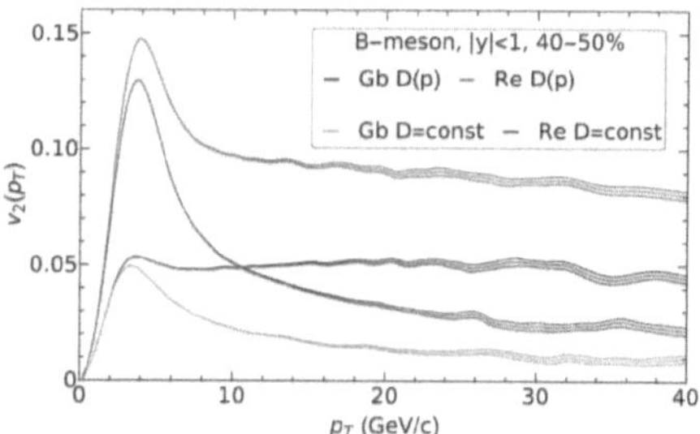

(a) B-meson $v_2(p_T)$ at $\sqrt{s_{NN}} = 5.5$ TeV for the 40-50% centrality class.

(b) Expanded view of the transverse momentum region, $0 < p_T \leq 30$ GeV/c of Fig. (a), on the left.

Figure 6.30: $v_2(p_T)$ for B-mesons at $\sqrt{s_{NN}} = 5.5$ TeV for both *Gb* and *Re* parameters with diffusion coefficients, $D = const$ as well as $D(p)$ at 40-50% centrality.

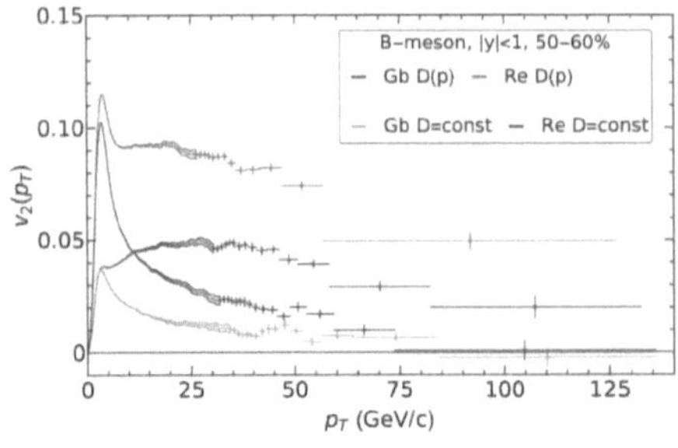

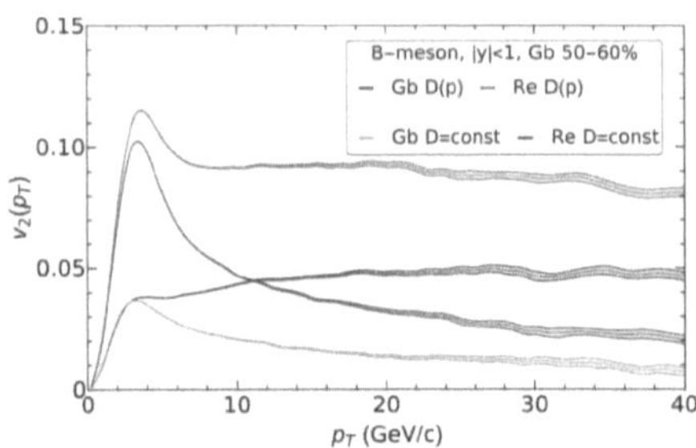

(a) B-meson $v_2(p_T)$ at $\sqrt{s_{NN}} = 5.5$ TeV for the 50-60% centrality class.

(b) Expanded view of the transverse momentum region, $0 < p_T \leq 30$ GeV/c of Fig. (a), on the left.

Figure 6.31: $v_2(p_T)$ for B-mesons at $\sqrt{s_{NN}} = 5.5$ TeV for both Gb and Re parameters with diffusion coefficients, $D = const$ as well as $D(p)$ at 50-60% centrality.

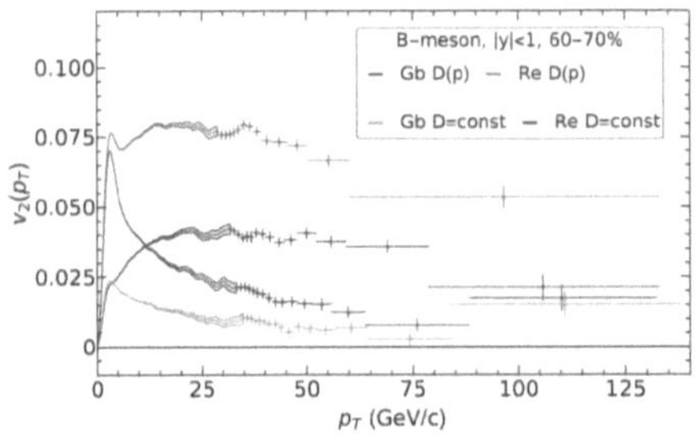
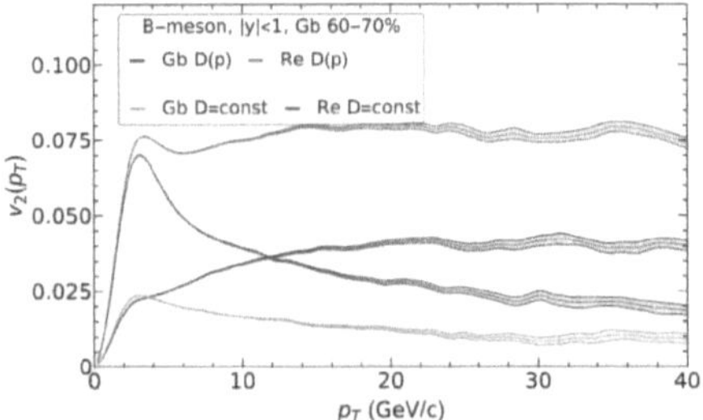

(a) B-meson $v_2(p_T)$ at $\sqrt{s_{NN}} = 5.5$ TeV for the 60-70% centrality class.

(b) Expanded view of the transverse momentum region, $0 < p_T \leq 30$ GeV/c of Fig. (a), on the left.

Figure 6.32: $v_2(p_T)$ for B-mesons at $\sqrt{s_{NN}} = 5.5$ TeV for both Gb and Re parameters with diffusion coefficients, $D = const$ as well as $D(p)$ at 60-70% centrality.

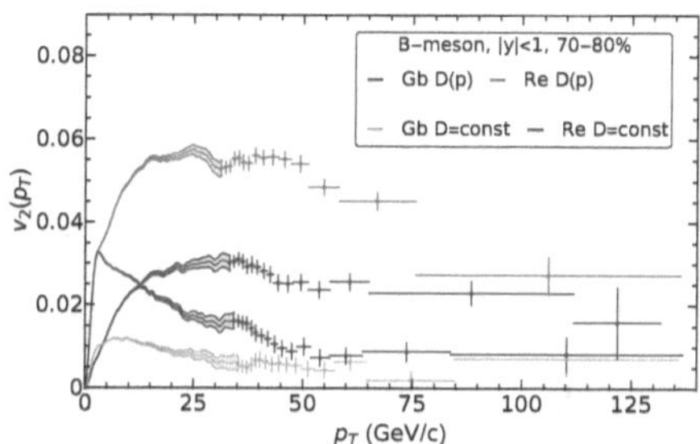
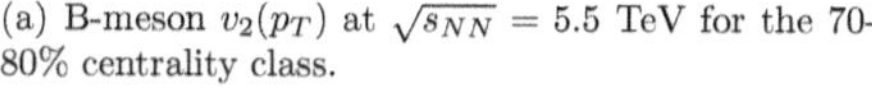

(a) B-meson $v_2(p_T)$ at $\sqrt{s_{NN}} = 5.5$ TeV for the 70-80% centrality class.

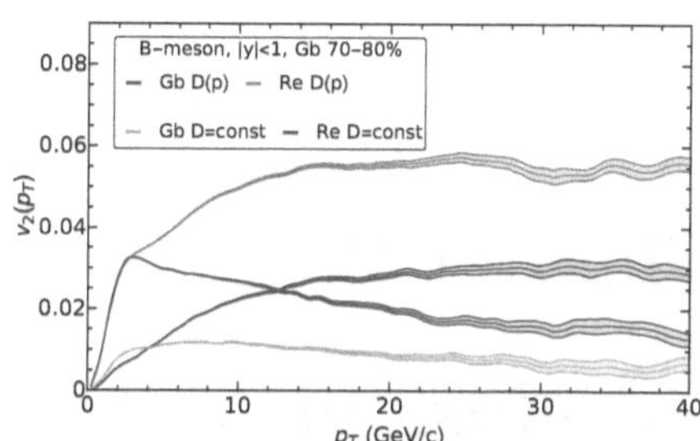

(b) Expanded view of the transverse momentum region, $0 < p_T \leq 30$ GeV/c of Fig. (a), on the left.

Figure 6.33: $v_2(p_T)$ for B-mesons at $\sqrt{s_{NN}} = 5.5$ TeV for both *Gb* and *Re* parameters with diffusion coefficients, $D = const$ as well as $D(p)$ at 70-80% centrality.

Chapter 7

Conclusions and Outlook

Heavy flavour energy loss is crucial for understanding properties of nuclear matter and the QCD phase diagram both theoretically and experimentally. One of the goals of this **book** is to improve our understanding of the properties of the QCD phase diagram at low baryon chemical potential and high temperatures (i.e. in the order of a few hundred MeV) which are experimentally accessible at RHIC and the LHC. We assumed strong coupling of the QGP and have used a Langevin energy loss model to study the suppression and angular distribu-tion of bottom quarks and consequently B-mesons at LHC energies. The Langevin energy loss model in this **book** is the first calculation of the suppression and azimuthal distribution of
B-mesons using AdS/CFT techniques [88, 113, 154] at $\sqrt{s_{NN}} = 2.76$ TeV and $\sqrt{s_{NN}} = 5.5$ TeV.

We've presented quantitative predictions for the $R_{AA}(p_T)$ and $v_2(p_T)$ for B-mesons at $\sqrt{s_{NN}} = 2.76$ TeV for central collisions and at $\sqrt{s_{NN}} = 5.5$ TeV for central, semi-central and peripheral collisions. These predictions have been made using different set of parameters, such as a diffusion coefficient that is dependent on momentum, $D(p)$ and a diffusion coefficient which does not depend on momentum, $D = const$. As shown in Eq. (4.7), momentum fluctuations in the longitudinal direction grow very quickly with the velocity of the heavy quark, (i.e. as $\gamma^{5/2}$). Therefore, predictions employing $D(p)$ break down at high momentum due to the strong contribution of longitudinal fluctuations on the energy loss. In addition to this, the translation of parameters from $\mathcal{N} = 4$ SYM theory to QCD is nontrivial. The translation of the drag coefficient given by Eq. (4.2) from $\mathcal{N} = 4$ SYM theory to QCD introduces some theoretical systematic uncertainties and to account for these, we've made predictions using two different sets of parameters given by Eq. (4.17) and Eq. (4.18).

We expect the nuclear modification factor to have a centrality dependence. The difference in suppression when moving from central to peripheral collisions can be seen from looking at our results in Fig. (4.2 - 4.5) where we compare the time spent by a heavy quark in the QGP medium depending on the centrality class it's produced in. We observe that a heavy quark produced at a particular point in the QGP in a central collision spends more time in the medium and consequently loses more energy compared to a heavy quark produced at the same point but in a peripheral collision. Our nuclear modification predictions are consistent with this observation as shown in Fig. (6.4 - 6.7) which show the $R_{AA}(p_T)$ dependence on centrality.

The nuclear modification factor results shown in Fig. (6.8 - 6.16) show that B-meson suppression is more pronounced in the intermediate momentum region, $5 < p_T < 25$ GeV/c. At low momentum, our predictions employing different parameters are qualitatively similar as the $R_{AA}(p_T)$ peaks, drops similarly and cross $R_{AA}(p_T) = 1$ at similar points. This is largely influenced by the bottom quark production spectrum which falls off following a power law dependence on momentum. The $R_{AA}(p_T)$ increases at high momentum but remains below one, a possible reason for the increase is that there are very few B-mesons produced with high p_T. At high transverse momentum, predictions employing $D(p)$ parameters show an over-suppression of B-mesons, for reasons that we've already discussed.

The $v_2(p_T)$ results for B-mesons at $\sqrt{s_{NN}} = 5.5$ TeV were presented in Sec. (6.2) and showed that $v_2(p_T)$ peaks at intermediate transverse momentum, $3 < p_T < 5$ GeV/c then decreases at high momentum. There is an anti-correlation between the $v_2(p_T)$ and the nuclear modification factor. The predicted $v_2(p_T)$ increases as we move from central collisions to mid-central collisions, then decreases as we move from mid-central collisions to peripheral collisions. This change in $v_2(p_T)$ with centrality can be attributed to the initial geometry of the overlap region such that $v_2(p_T)$ is low in central collisions since the overlap region is almost symmetrical, so the low spatial anisotropy converts to a low azimuthal anisotropy. The maximum $v_2(p_T)$ is obtained in mid-central collisions where the spatial asymmetry of the initial collision geometry is at its largest.

Comparing theoretical models to experimental data is crucial and allows for the possibility of falsifying some models. We've shown in Fig. (6.2) for the nuclear modification factor and Fig. (6.17) for the $v_2(p_T)$ that our predictions at $\sqrt{s_{NN}} = 2.76$ TeV are qualitatively consistent with nonprompt J/Ψ measured by CMS [73] shown in Fig. (6.3) and Fig. (6.18) respectively. These qualitative comparisons made with CMS data have limitations since the CMS data is for nonprompt J/Ψ and not B-mesons, the CMS data also covers a larger rapidity range and a wider centrality class. These limitations have been discussed in Sec. (6.1) for $R_{AA}(p_T)$ and Sec. (6.2) for $v_2(p_T)$. We've also provided falsifiable $R_{AA}(p_T)$ and $v_2(p_T)$ predictions for future measurements at the LHC run 3 expected to start in 2021 [155].

In this book, we've performed phenomenological calculations for B-mesons that can be compared to data. One can also perform these calculations for D-mesons as well as other collision systems such as $Xe + Xe$ [156] and this is left for future work. It would also be valuable to investigate whether the theoretical framework of AdS/CFT can be improved. In particular, the mapping between QCD and $\mathcal{N} = 4$ SYM theory. Our energy loss model only considers heavy quark propagation beyond thermalisation (taken to be at $\tau \sim 0.6$ fm/c), incorporating pre-thermalisation energy loss effects could provide insight on the motion of the heavy quark prior the applicability of hydrodynamics. Since the initial production of high-p_T particles is described by pQCD, using a pQCD energy loss model before thermalisation followed by a strong coupling treatment post thermalisation may be a reasonable approach. One could also investigate whether AdS/CFT energy loss calculations can be applied to low energy heavy-ion collisions, this may require one to account for the non-zero baryon chemical potential and low temperature effects on the drag and diffusion terms.

Appendix A

Generating Random Numbers

A.1 Cumulative probability distribution functions

In order to produce random numbers obeying any distribution, we need to have the Probability Distribution Function (PDF) which can either be given as an analytic expression or numerically. For a given PDF, we can compute its corresponding Cumulative Distribution Function (CDF) by integrating along the real line. A CDF is a distribution function of a continuous random variable (X) that is evaluated at x and it gives the probability that X will take a value less than or equal to x. This is shown in Eq (A.1), where ρ is the PDF.

$$P_X(x) \;=\; \int_{-\infty}^{x} \rho_X(t)dt \tag{A.1}$$

The CDF has the properties that it is non-decreasing (its derivative is a density function which can not be negative) and right-continuous with:

$$\lim_{x \to -\infty} P(x) \;=\; 0 \tag{A.2}$$

$$\lim_{x \to \infty} P(x) \;=\; 1 \tag{A.3}$$

In making use of cumulative distribution functions to generate random numbers, we'll consider two cases: one where the PDF is defined by an analytic expression and the other where it is only defined numerically.

PDF defined analytically
Suppose that we have a PDF defined by the following exponential function:

$$\rho(t) \;=\; \beta e^{-\beta t} \,, t > 0 \tag{A.4}$$

We compute the corresponding CDF by integrating this PDF within its domain as follows:

$$
\begin{aligned}
P(x) &= \int_{-\infty}^{x} \rho(t)dt & \text{(A.5)}\\
&= \beta \int_{-\infty}^{x} e^{-\beta t} & \text{(A.6)}\\
&= 1 - e^{-\beta x} \ , \ x > 0 & \text{(A.7)}\\
&= CDF(x) & \text{(A.8)}
\end{aligned}
$$

Note that for the considered PDF, the CDF is defined for all $x > 0$ and gives the area under the exponential PDF within the interval $(0, x]$ where $x \in \mathbb{R}^{+}$. The inverse function of the CDF (inverse CDF) can also be computed analytically as follows:

$$
\begin{aligned}
x &= 1 - e^{-\beta y} & \text{(A.9)}\\
e^{-\beta y} &= 1 - x & \text{(A.10)}\\
-\beta y &= ln(1 - x) & \text{(A.11)}\\
y &= \frac{-1}{\beta} ln(1 - x) & \text{(A.12)}\\
P^{-1}(x) &= \frac{-1}{\beta} ln(1 - x) & \text{(A.13)}\\
&= InvCDF(x) & \text{(A.14)}
\end{aligned}
$$

An important thing to note in our example of the exponential PDF is that, the domain of the CDF is $(0, x]$ and its range is $(0, 1]$. While the inverse CDF has a domain of $(0, 1]$ and a range of $(0, x]$. It is true in general that the domain of the CDF is the range of its inverse and vice-versa.

PDF defined only numerically

Most distributions (PDFs) that we will deal with in Physics (and particularly in this **book**) are only defined numerically, that is, we only have a set of coordinates (x, y) for the density function (which is mostly continuous) at discrete points within its domain. As a result, we can't compute its corresponding inverse CDF analytically and have to resort to numerical methods. To obtain the CDF, we integrate the given discrete PDF coordinates using one of the various numerical integration schemes, i.e trapezoidal rule [157]. The inverse CDF is then found by swapping the domain and the range of the CDF.

Since the ultimate goal is to generate random numbers obeying a given probability density function, for analytically defined functions; once we have the inverse CDF, we evaluate it at random points, $x \in (0, 1)$ using the various schemes described in Ref. [120]. This will give us random numbers obeying the respective PDF. For functions defined only numerically; one needs to interpolate the inverse CDF and evaluate it at those points to generate random numbers obeying the respective numerical PDF. In this work, we made use of cubic spline interpolation for single variable functions (i.e when producing quarks along a line) and bilinear interpolation for functions of two variables (i.e when producing quarks in the transverse plane). For a more extensive discussion on random number generation, see Ref. [120].

Appendix B

Relativistic Heavy-ion Collisions and Hydrodynamics

B.1 Hydrodynamic description of ultrarelativistic heavy-ion collisions

B.1.1 Hydrodynamics and classical fluids

Hydrodynamics is a classical effective field theory (neglects relativistic or quantum mechanical effects) it describes a continuous medium such as a fluid or gas by making use of averages over the system's microscopic degrees of freedom of the system and assigning physical quantities a certain value at every point in space and time. To reduce the degrees of freedom in the physical system, coarse graining is used, new degrees of freedom are chosen by taking an average over all the particles in a small volume (V), often referred to as the fluid-particle with infinitesimal elements d^3V, called material points in space (geometrical manifold) and choose that volume as the new fundamental degrees of freedom of the system. The fluid particle (or cell) is defined by its position $(\vec{r})$ and parameters such as internal energy, number of atoms/particles in it etc. Even if the global system is isolated, energy or matter can be transported from one fluid particle to another and generally, the local extensive variables are time-dependent. The fluid-particle needs to obey two conditions;

- it has to be larger than the real particles (i.e elementary particles) of the system such that it can be meaningfully treated as a thermodynamic system. This ensures that the relative microscopic fluctuations of thermodynamic quantities computed for the fluid particle are negligible.

- it has to be much smaller than the full (bigger) system that needs to be described, such that it acts as point-like particles in the full system. This ensures that homogeneity is approximately ensured and thermodynamic properties can vary slightly from cell (fluid particle) to cell.

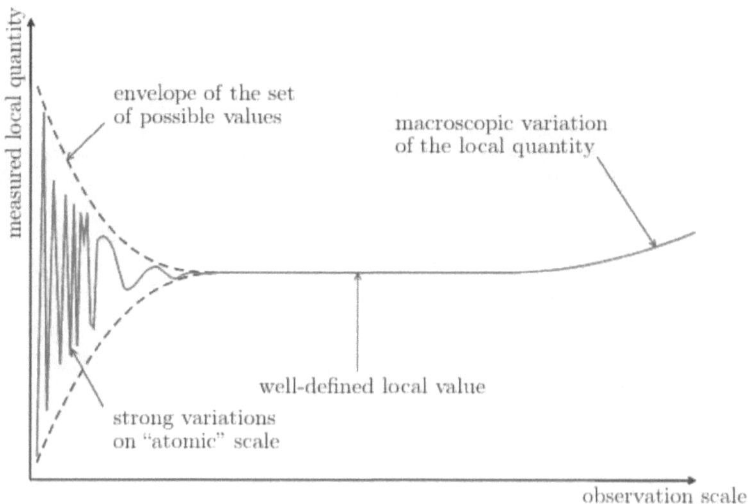

Figure B.1: Typical variation of the value of a 'local' macroscopic observable measured at different scales [158]

The importance of these two conditions is depicted in Fig. (B.3), showing a schematic view of how the measured value of a local macroscopic quantity (i.e density) varies as a function of the length scale in which it is defined. If the resolution of the measurement is too high (i.e small length scale), then the discrete degrees of freedom of the system become relevant and the value we measure fluctuates strongly from one observation point to the next (statistical fluctuations), as shown by enveloped results in Fig. (B.3). The fluctuating behaviour of the measurement decreases as the observation scale is increased, since this results in an increase in the number of atoms/particles inside the probed volume (fluid particle). This is particularly addressed by the first condition. On the other hand, as the observation scale is increased, we reach a point where the measurement resolution is low, i.e the fluid particle has enough atoms/particles to overcome statistical fluctuations but inhomogeneous macroscopic properties kick in. This would violate the second condition. Therefore, the favourable regime (which doesn't necessarily exist in all systems) to make the measurement of local macroscopic quantities is the region in-between the two scales defined above, where the observable doesn't depend on the scale in which it's being measured.

If these two conditions are satisfied for a particular system, we can define local thermodynamic variables, which correspond to the values measured in the fluid particle. Given that the actual physical size of each fluid cell is irrelevant, there is no meaningful local variable corresponding to the fluid particle (volume) and the values of the extensive variables are also arbitrary. We replace them by local densities such as the number density $n(t, \vec{r})$, energy density $e(t, \vec{r})$, entropy density $s(t, \vec{r})$, mass density $\rho(t, \vec{r})$, linear momentum density $\rho(t, \vec{r})\vec{v}(t, \vec{r})$ etc. In the assumed local thermal equilibrium, the equation(s) of state of the fluid particle system (defined using local thermodynamic quantities), is the same as that of a macroscopic system in the thermodynamic limit of infinitely large volume and large particle number.

As an example, we can take a non-relativistic classical ideal gas with the equation of state given by Eq. (B.1),

$$PV \;=\; Nk_BT \tag{B.1}$$

describing a gas with N number of atoms, occupying a volume V at constant pressure P and temperature T. This equation is recast to Eq. (B.2),

$$P(t, \vec{r}) \;=\; n(t, \vec{r})k_BT(t, \vec{r}) \tag{B.2}$$

which is the equation of state for a local fluid particle (at position $\vec{r}$), where n is the number density of the atoms, $T(t, \vec{r})$ is the local temperature, $P(t, \vec{r})$ is the local pressure and k_B is the Boltzmann constant. This equation now defines the assumed local thermodynamic equilibrium under non-uniform temperature and pressure. The vector $\vec{r}$ is promoted to a continuous variable in $\mathbb{R}^3$ (or specifically in the volume at an instant t, V_t) while the thermodynamic parameters become fields on $\mathbb{R} \times \mathbb{R}^3$.

In addition, systems that satisfy these two conditions fulfil the Knudsen-number relation [159] given in Eq. (B.3).

$$K_n \;\equiv\; \frac{l_{mfp}}{L} \ll 1 \tag{B.3}$$

$$L \;\cong\; \left[\frac{|\vec{\nabla}G(t, \vec{r})|}{|G(t, \vec{r})|} \right]^{-1} \tag{B.4}$$

where l_{mfp} is the mean free path (i.e the distance travelled by a particle before it interacts with another particle) of the particles of the original system while L is the typical size of the system (which is the length scale at which the macroscopic physical properties of the system may vary) and these give the two length scales of interest. The parameter G denotes the macroscopic physical quantity under consideration and $\vec{\nabla}$ is the spatial gradient. Particles in systems that obey Eq. (B.3) will almost certainly interact with each other. The regime where $K_n > 1$, defines what is known as the Knudsen gas [160] where any collisions between atoms/particles is negligible i.e insufficient to ensure that thermal equilibrium is established as in the case of an ideal gas. In this regime, hydrodynamics doesn't hold and descriptions such as molecular dynamics [161–163] are utilised.

There are two main descriptions used to describe the evolving system, the Lagrangian formalism as well as the Eulerian formalism. The Lagrangian formalism focuses on the trajectories of material points, defined by their position, $\vec{r} = \vec{r}(t, \vec{R})$ in the reference configuration ($\vec{R} = X^i \vec{e_i}$) with the consistency condition $\vec{r}(t = t_0, \vec{R}) = \vec{R}$. Generalising this, we can describe the various physical quantities (G) of a continuous medium as $G = G(t, \vec{R})$. The velocity, $\vec{v}(t, \vec{r})$, and acceleration, $\vec{a}(t, \vec{r})$, of the material point can be determined by taking partial derivatives with respect to time. The Eulerian formalism is commonly used in fluid dynamics, it focuses on geometrical points and uses the system configuration at time t as the reference point for the evolution between instants t and $t + dt$. The physical quantities are described by fields on spacetime and the velocity field, $\vec{v_t}(t, \vec{r})$ is the fundamental field that determines the motion of a continuous medium. The velocity field is defined as the Lagrangian velocity, $\vec{v} = \vec{v_t}(t, \vec{r})$, of a material point passing through $\vec{r}$ at time t and any physical quantity G is given by the function $G = G_t(t, \vec{r})$.

We can consider a material point in a continuous medium with the following position vectors at successive instants $(\vec{r}, t)$ and $(\vec{r} + d\vec{r}, t + dt)$, as well as velocity $\vec{v}(t, \vec{r})$ and $\vec{v}(t + dt, \vec{r} + d\vec{r})$. The displacement of the material point between these successive points is given by $d\vec{r} = \vec{v}(t, \vec{r})dt$, for small dt. If we define $d\vec{v} \equiv \vec{v}(t + dt, \vec{r} + d\vec{r}) - \vec{v}(t, \vec{r})$, assuming $\vec{v}(t, \vec{r})$ is differentiable, the lowest order Taylor expansion of $d\vec{v}$ is given in Eq. (B.5),

$$d\vec{v} \cong \frac{\partial \vec{v}(t, \vec{r})}{\partial t}dt + \frac{\partial \vec{v}(t, \vec{r})}{\partial x^1}dx^1 + \frac{\partial \vec{v}(t, \vec{r})}{\partial x^2}dx^2 + \frac{\partial \vec{v}(t, \vec{r})}{\partial x^3}dx^3 \tag{B.5}$$

$$= \frac{\partial \vec{v}(t, \vec{r})}{\partial t}dt + (d\vec{r} \cdot \vec{\nabla})\vec{v}(t, \vec{r}) \tag{B.6}$$

and the acceleration is given by Eq. (B.7),

$$\vec{a}(t) = \frac{\partial \vec{v}(t, \vec{r})}{\partial t} + [\vec{v}(t, \vec{r}) \cdot \vec{\nabla}]\vec{v}(t, \vec{r}) \tag{B.7}$$

$$= \frac{D\vec{v}(t, \vec{r})}{Dt} \tag{B.8}$$

where the two operators used, $d\vec{r} \cdot \vec{\nabla}$ and D/Dt (known as the hydrodynamic derivative) are defined by equations Eq. (B.9) and Eq. (B.10) respectively.

$$d\vec{r} \cdot \vec{\nabla} \equiv dx^1\frac{\partial}{\partial x^1} + dx^2\frac{\partial}{\partial x^2} + dx^3\frac{\partial}{\partial x^3} \tag{B.9}$$

$$\frac{D}{Dt} \equiv \frac{\partial}{\partial t} + \vec{v}(t, \vec{r}) \cdot \vec{\nabla} \tag{B.10}$$

The first term in Eq. (B.7) is the local acceleration and follows from the non-stationarity of the velocity field while the second term is the convective acceleration due to the non-uniformity of the motion.

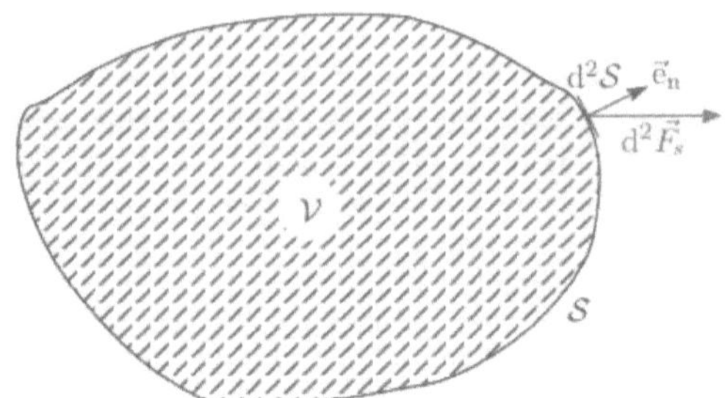

Figure B.2: Illustration of a surface force $d^2\vec{F}_s$ acting on the surface S of an infinitesimally small volume V [158]

There are two types of forces that act on a closed material domain V that lies inside the volume V_t occupied by a continuous medium. Volume/body forces (i.e weight or long-range electromagnetic force), which act on each point of the bulk volume of V and surface/contact forces (i.e friction) which act on the surface S, where S is the geometric surface enclosing V. One can look at an infinitesimally small geometrical surface element d^2S at a point in V, experiencing a surface (contact) force $d^2\vec{F}_s$ that is caused by the medium outside V as shown in Fig. (B.4). The surface density of the contact forces is given by Eq. (B.11),

$$\vec{T}_s = \frac{d^2\vec{F}_s}{d^2S} \tag{B.11}$$

known as the mechanical stress vector on d^2S. It can be decomposed into normal stress when it's directed towards the interior of V (i.e orthogonal to the tangent plane) and shear stress when it acts in the tangent plane. A fluid is described as a continuous medium that deforms when it experiences shear stresses (i.e it doesn't support shear stress) and its motion is described by hydrodynamics.

Consider a Newtonian fluid, a fluid that has a constant viscosity and whose shear rate is directly proportional to the shear stress, such as water or air. Its strain tensor is given by Eq. (B.12)

$$e_{ij} = \frac{1}{2}\left(\frac{\partial v_i}{\partial x^j} - \frac{\partial v_j}{\partial x^i}\right) \tag{B.12}$$

where v_i are the ith components of the fluid velocity and x^i are the components of the coordinate vectors. The fluid needs to satisfy four equations:

- Continuity equation, which is derived from the conservation of mass and given by Eq. (B.13),

$$\frac{D\rho}{Dt} \;=\; -\rho\vec{\nabla}\cdot\vec{v} \tag{B.13}$$

  where ρ is the density of the fluid.

- Momentum equation (i.e momentum conservation), given by Eq. (B.14),

$$\frac{Dv_j}{Dt} \;=\; \frac{\partial T_{ij}}{\partial x^j} - \rho g_j \tag{B.14}$$

$$T_{ij} \;=\; -P\delta_{ij} + 2\mu e_{ij} + \lambda e_{mm}\delta_{ij} \tag{B.15}$$

  where T_{ij} is the stress tensor, P is the pressure, g_j is the gravitational acceleration, δ_{ij} is the unity matrix, μ and λ are fluid dependent scalars. If the fluid is incompressible ($\vec{\nabla}\cdot\vec{v} = 0$), this reduces to the Navier-Stokes Equations.

- Equation of state, which gives the relationship between the pressure (P), temperature (T), as well as density (ρ). It varies based on the fluid under consideration and is given by Eq. (B.2) for an ideal gas.

- Temperature/Energy equation, which deals with the thermodynamic effects within the medium. If we consider the heat flux vector q_i at any given point, we need to solve Eq. (B.16)

$$\rho\frac{De}{Dt} \;=\; \frac{\partial q_i}{\partial x_i} - P\left(\frac{\partial v_i}{\partial x_i}\right) + \phi \tag{B.16}$$

  for the internal energy (e), where ϕ is the viscous dissipation. The temperature/energy equation indicates that any change in energy in the fluid is due to convergence of heat, volume compression and viscous dissipation.

B.1.2   Relativistic hydrodynamics

A relativistic fluid is basically a classical fluid modified by the laws of special relativity as well as curved spacetime (i.e general relativity). In relativistic heavy-ion collisions, particles in the quark-gluon plasma may have a very high kinetic energy as compared to their rest energy. This results in the energy density of the fluid being very large such that a non-relativistic description of hydrodynamics can't be applied. Large systems that contain a large number (order of Avogadro's number) microscopic parts have small fluctuations and reach a state of maximum disorder rapidly due to microscopic dynamics, where their global behavior can be described by few macroscopic thermodynamic fields. Systems produced in relativistic heavy-ion collisions are much smaller, and contain particles on the order of several thousands. A hydrodynamic description of systems of atomic gases has been studied in Ref. [164, 165].

The hydrodynamic approach is independent on the total number of particles in the system, but requires momentum transfer rates that are occurring on a microscopic level to be sufficiently large such that the system rapidly relaxes to a local thermal equilibrium configuration on macroscopic time scales [166]. Due to the high energies involved in a relativistic fluid, the particle number is not conserved since particle-antiparticle pairs are created or annihilated. Instead, if particles carry a conserved quantum number (i.e electric charge), the quantum number difference between particles and antiparticles is conserved. We can consider a relativistic fluid with a single species of particles for simplicity, one can define the local particle number (defined as the difference between particles and antiparticles) density, $n(t, \vec{r})$ such that $n(t, \vec{r}) \cdot d^3\vec{r}$ gives the total number of particles at the point $\vec{r}$ at time t, where $d^3\vec{r}$ is the infinitesimal spatial volume element.

Perfect fluid

We now discuss a perfect (ideal) fluid, which is a fluid with no dissipative currents (i.e friction and heat flow). This allows us to define a reference frame (at each fluids' spacetime point) such that the local surrounding of the particular point is spatially isotropic [50] and this reference frame is usually adopted as the local rest frame. The particle flux density $\vec{j}_N(t, \vec{r})$ is the number of particles crossing a unit surface per unit time interval. The local particle number density and the particle flux density when combined, they give the particle number four-current $N(t, \vec{r})$ given by Eq. (B.17)

$$N^\mu(t, \vec{r}) \;=\; \begin{pmatrix} n(t, \vec{r}) \\ \vec{j}_N(t, \vec{r}) \end{pmatrix} \tag{B.17}$$

and the local formulation of particle number conservation is given by Eq. (B.18).

$$\partial_\mu N^\mu \;=\; 0 \tag{B.18}$$

The energy momentum tensor, $T^{\mu\nu}_{rest}(t, \vec{r})$ [167] describes the conservation of the four-momentum locally in terms of densities as well as flux densities of the energy and the momentum at a particular point in spacetime. If we have a thermalised fluid cell, its energy momentum tensor in its local rest frame is given by Eq. (B.19),

$$T^{\mu\nu}_{rest}(x) \;=\; diag(\epsilon(x), P(x), P(x), P(x)) \tag{B.19}$$

where we've replaced $(t, \vec{r})$ by x representing the position of the fluid cell with energy density $\epsilon(x)$ and pressure $P(x)$. Suppose the fluid cell is moving with a four-velocity $u^\mu(x)$ in a global reference frame, given by Eq. (B.20),

$$u^\mu(x) \;=\; \gamma(1, v_x, v_y, v_z), \quad \gamma = \frac{1}{\sqrt{1 - v^2}} \tag{B.20}$$

$$u^\mu(x)u_\mu(x) \;=\; 1, \quad \forall x \tag{B.21}$$

a boost of $T^{\mu\nu}_{rest}(x)$ gives the energy momentum tensor of the fluid in the global frame given by Eq. (B.22), with particle number four-current in Eq. (B.23).

$$
\begin{aligned}
T^{\mu\nu}(x) &= (\epsilon(x) + P(x))u^{\mu}(x)u^{\nu}(x) - P(x)g^{\mu\nu}(x) & \text{(B.22)} \\
N^{\mu}(x) &= n(x)u^{\mu}(x) & \text{(B.23)}
\end{aligned}
$$

The components of the energy momentum tensor are defined as follows:

- $T^{00}(x)$ is the energy density

- $T^{0j}(x)$ is the the jth component of the energy flux density, for $j = 1, 2, 3$

- $T^{i0}(x)$ is the density of the ith momentum component, for $i = 1, 2, 3$

- $T^{ij}(x)$ is the momentum flux-density tensor, for $i, j = 1, 2, 3$

The energy-momentum tensor can be rewritten as Eq. (B.24)

$$
\begin{aligned}
T^{\mu\nu}(x) &= \epsilon(x)u^{\mu}(x)u^{\nu}(x) + P(x)\Delta^{\mu\nu}(x) & \text{(B.24)} \\
\Delta^{\mu\nu}(x) &= u^{\mu}(x)u^{\nu}(x) - g^{\mu\nu}(x) & \text{(B.25)}
\end{aligned}
$$

by introducing a tensor, $\Delta^{\mu\nu}(x)$, that is a projector on the 3D vector space that is orthogonal to the four-velocity. The local conservation of the energy-momentum tensor is given by Eq. (B.26).

$$
\partial_{\mu}T^{\mu\nu} = 0, \quad \nu = 0, 1, 2, 3 \tag{B.26}
$$

The local conservation of the energy-momentum tensor gives the conservation of energy for $\nu = 0$ and conservation of momentum in each component for $\nu = 1, 2, 3$. In the fluid's local rest frame, the spatial components of the four-velocity vanish as shown in Eq. (B.27).

$$
u^{\mu}(x) = (1, 0, 0, 0) \tag{B.27}
$$

If $\vec{v}(x)$ denotes the instantaneous velocity of an observer at rest in the fluid's local rest frame, the components of the four-velocity in the fixed reference frame are given by Eq. (B.28).

$$
u^{\mu}(x) = \begin{pmatrix} \gamma(x) \\ \gamma(x)\vec{v}(x) \end{pmatrix} \tag{B.28}
$$

In the local rest frame, local thermodynamic variables are assumed to be related to the particle number density and the energy density as is the case when the fluid is at thermodynamic equilibrium. The particle number four-current, $N^{\mu}(x)$ reduces to Eq. (B.29) and Eq. (B.30).

$$
\begin{aligned}
N^{0}(x) &= n(x) & \text{(B.29)} \\
N^{i}(x) &= 0, \quad i = 1, 2, 3 & \text{(B.30)}
\end{aligned}
$$

Dissipative fluid

As discussed earlier, the hydrodynamic description requires the system to establish a local thermal equilibrium very fast. In the event where this is not the case (i.e the local relaxation rates are not fast to ensure a rapid local thermalization), dissipative terms that are proportional to transport coefficients of bulk and shear viscosity, heat conduction and diffusion need to be included in the energy momentum tensor as well as particle number four-current definitions [166–169]. The introduction of the terms breaks the local isotropy of the fluid, leading in the inability to uniquely define the fluid's local rest frame. The dissipative effects are as a result of spatial gradients of the flow velocity field, the temperature, or the chemical potential that is associated with the conserved particle number. The equivalent description of the particle number four-current and energy-momentum tensor in the dissipative case are given by Eq. (B.31) and Eq. (B.32) respectively.

$$N^\mu(x) \;=\; N_0^\mu(x) + n^\mu(x) \tag{B.31}$$
$$T^{\mu\nu} \;=\; T_0^{\mu\nu} + \tau^{\mu\nu} \tag{B.32}$$

where quantities with subscript 0 correspond to the perfect fluid case, $n^\mu(x)$ are the components of a four-vector representing the dissipative particle number four-current flux density and $\tau^{\mu\nu}$ are components of a rank two tensor representing dissipative energy-momentum flux density. The solution to these generalised equations is discussed in Ref. [170]. One introduces a four-velocity, $u^\mu(x)$, that obeys Eq. (B.21) such that the reference frame in which the spatial components of $u^\mu(x)$ vanish will be adopted as the local rest frame that is associated with $u^\mu(x)$. The projector, $\Delta^{\mu\nu}(x)$, that is orthogonal to the four-velocity on the three-dimensional vector space is defined as in Eq. (B.25). This allows us to rewrite the particle number four-current, Eq. (B.33),

$$N^\mu(x) \;=\; n(x)u^\mu(x) + n^\mu(x) \tag{B.33}$$

as well as the energy-momentum tensor, Eq. (B.34), in the dissipative case.

$$T^{\mu\nu}(x) \;=\; (\epsilon(x) + P(x))u^\mu(x)u^\nu(x) - P(x)g^{\mu\nu}(x) + \tau^{\mu\nu}(x) \tag{B.34}$$

The condition in Eq. (B.35)

$$u_\mu(x)n^\mu(x) \;=\; 0 \tag{B.35}$$

holds in any coordinate system since the left hand side is a Lorentz scalar. The quantity $n^\mu(x)$ represents a particle number four-current that is diffusive in the local rest frame and describes the dissipative transport of the particle number. The dissipative energy-momentum term $\tau^{\mu\nu}(x)$ doesn't have a 00-component in the local rest frame such that $T^{00}(x)$ still defines the energy density, $\epsilon(x)$ in that frame. The most general rank two symmetric tensor which obeys this condition has the form in Eq. (B.36).

$$\tau^{\mu\nu}(x) \;=\; q^\mu(x)u^\nu(x) + q^\nu(x)u^\mu(x) + \pi^{\mu\nu}(x) \tag{B.36}$$

where $q^\mu(x)$ are the four-vector components and $\pi^{\mu\nu}(x)$ are components of a rank two tensor such that the following conditions hold:

$$u_\mu(x)q^\mu(x) \;=\; 0 \tag{B.37}$$
$$u_\mu(x)\pi^{\mu\nu}(x)u_\nu(x) \;=\; 0 \tag{B.38}$$

The quantity $q^\mu(x)$ is the heat current or energy flux current in the fluid's local rest frame and the condition in Eq. (B.37) says that it needs to be orthogonal to the four-velocity. On the other hand, $\pi^{\mu\nu}(x)$ is known as the stress tensor and contains all dissipative phenomena due to friction forces. It is a symmetric tensor that can be broken down into the sum of a traceless tensor $\omega^{\mu\nu}(x)$ and a tensor proportional to the projector given in Eq. (B.25). The decomposition is as follows:

$$\pi^{\mu\nu}(x) \;=\; \omega^{\mu\nu}(x) + \Pi(x)\Delta^{\mu\nu}(x) \tag{B.39}$$

The traceless tensor $\omega^{\mu\nu}(x)$ is the shear stress tensor in the fluid's local rest frame, it describes the transport of momentum as a result of shear deformations. The $\Pi(x)$ is a dissipative pressure term as it behaves like the thermodynamic pressure, $P(x)$. The energy-momentum tensor of a dissipative relativistic fluid given in Eq. (B.34) can now be rewritten as follows:

$$T^{\mu\nu}(x) \;=\; \epsilon(x)u^\mu(x)u^\nu(x) + [P(x) + \Pi(x)]\Delta^{\mu\nu}(x) \tag{B.40}$$
$$+ \; q^\mu(x)u^\nu(x) + q^\nu(x)u^\mu(x) + \omega^{\mu\nu}(x) \tag{B.41}$$

At any given point in a dissipative fluid, the energy as well as the particle number may not necessarily be flowing in the same direction. This is a result of the energy that may still be transported by particle-antiparticle pairs that don't make contributions to the particle density. Possibly, it's also because of the various conserved quantum numbers that flow in different directions. Generally, one can't uniquely pick the four-velocity of the fluid flow since it's impossible to find a reference frame that is preferable such that the local properties of the fluid are isotropic. Several definitions of the four-velocity have been given and these give different descriptions of the dissipative currents, but the physics remains the same for all the different definitions. One choice of the local rest frame is the Eckart frame [171], where the four-velocity is taken to be proportional to the particle number four-current and the dissipative particle number flux, $n^\mu(x) \to 0$. Another local rest frame choice is the Landau frame [172], where the four-velocity is taken to be proportional to the energy flux density, consequently, the heat current $q^\mu(x) \to 0$ and the dissipative tensor, $\tau^{\mu\nu}(x)$, is reduced to its viscous part, $\pi^{\mu\nu}(x)$.

To describe the physics of the fluid, we need the equation of state which relates the energy density, $\epsilon(x)$ to the pressure $P(x)$ and particle number $n(x)$ as well as an equation that describes the dissipative effects such as the diffusive particle number four-current $n^\mu(x)$, the heat flux density $q^\mu(x)$ etc. Theories of first order dissipative fluid dynamics [173–175] only include first order terms in the derivatives of quantities such as the temperature, the velocity or the chemical potential and they are relativistic generalisations of the laws that are valid for classical dissipative fluids. The disadvantage of these first order theories is that many solutions to their equations are not stable under small perturbations and the disturbances grow exponentially with respect to time on microscopic timescales. This in turn, could result in the velocity of certain modes exceeding the speed of light [176], thus breaking relativistic theory. In addition, the small gradient assumption used in first order dissipative fluid dynamics also breaks down since gradients grow quickly [177]. In order to do accurate computations in dissipative relativistic fluid dynamics, it is neccessary to go beyond a first order expansion in gradients, i.e second order as described by the Israel-Stewart-theory [178].

In first order, in the Landau frame we get the following equations for the dissipative pressure, stress tensor and particle number four-current:

$$\Pi(x) \;=\; -\zeta(x)\nabla^{\mu}(x)u_{\mu}(x) \tag{B.42}$$

$$\omega^{\mu\nu}(x) \;=\; -\eta(x)\left[\nabla^{\mu}(x)u^{\nu}(x) + \nabla^{\nu}(x)u^{\mu}(x) - \frac{2}{3}[\nabla^{\nu}(x)u_{\nu}(x)]\right] \tag{B.43}$$

$$n^{\mu}(x) \;=\; \kappa(x)\left[\frac{n(x)T(x)}{\epsilon(x)+P(x)}\right]^{2}\nabla^{\mu}(x)\left[\frac{\mu(x)}{T(x)}\right] \tag{B.44}$$

$$\nabla^{\mu}(x) \;\equiv\; \Delta^{\mu\nu}(x)d_{\nu} \tag{B.45}$$

The coefficient ζ describes the bulk viscosity, η gives the shear viscosity and κ describes the heat conductivity, all named after the dissipative phenomenon that they contribute to. They are all positive numbers that implicitly depend on the spacetime position and vary with temperature and chemical potential.

Appendix C

Parameters of the Langevin Energy Loss Model

As the heavy quarks move through the quark gluon plasma (QGP), they frequently experience soft momentum kicks from partons in the medium and consequently resemble the behaviour of Brownian motion. The dynamics of a heavy quark propagating through the QGP can be described by both the Boltzmann and Langevin models [179, 180]. In the Boltzmann approach, the Boltzmann Transport Equation (BTE) gives the distribution function describing the heavy quark evolution and a scattering matrix is used to quantify both elastic and inelastic processes that the heavy quark experiences as it traverses the QGP. The typical momentum transfers in the interactions are assumed to be small, $q \sim gT \ll m_Q$ [181], and consequently, the trajectory of the heavy quark only changes after it has received a lot of soft momentum kicks, resulting in Brownian motion. This assumption allows the Boltzmann Transport Equation to be reduced to the Fokker-Plank Transport Equation (FPTE), which helps reduce the problem to finding three transport coefficients that can be extracted from lattice QCD in the zero momentum limit [182]. The Boltzmann Transport Equation is given by Eq. (C.1):

$$\frac{p_Q}{E_Q} \cdot \partial f_Q \;=\; C[f_Q] \tag{C.1}$$

where p_Q is the heavy quark four-momentum, E_Q its energy, f_Q is the distribution function and $C[f_Q]$ is the collision integral which includes all interaction mechanisms between the heavy quarks and other constituent partons in the QGP. The Boltzmann Transport Equation can be linearised by ignoring the thermal parton distribution change in the medium due to the heavy quark propagation and the collision integral $C[f_Q]$ reduces to a linear function of f_Q. Solving Eq. (C.1) is not trivial, some Monte Carlo techniques can be used to solve it numerically, more extensive discussions can be found in refs. [182–186]. The Langevin approach used in this paper is discussed in Chap. (4) and its parameters are discussed in detail in this appendix.

C.1 AdS/CFT and $\mathcal{N} = 4$ Supersymmetric Yang-Mills theory

The Anti-de Sitter/Conformal Field Theory (AdS/CFT) correspondence [23–26, 187, 188] gives a mapping between $\mathcal{N} = 4$ $SU(N_c)$ supersymmetric Yang-Mills theory (SYM) and type IIB string theory in a $AdS_5 \times S^5$ gravitational background, where AdS_5 refers to the five dimensional Anti-de Sitter space and S^5 is a five dimensional sphere. This was strongly motivated by the discovery that there exists a string dual of QCD (or any gauge theory), which came from considering the limit of large 't Hooft's coupling, $\lambda = g_{YM}^2 N_c \gg 1$ in the large number of colours limit, $N_c \to \infty$ [90, 189, 190]. In $\mathcal{N} = 4$ SYM, the quarks are fundamental representation particles introduced by the addition of an $\mathcal{N} = 2$ hypermultiplet that has arbitrary mass, meaning adding a Dirac fermion and two complex scalars (in the fundamental representation), that have a common mass and Yukawa interactions that preserve $\mathcal{N} = 2$ supersymmetry [90]. These quarks can be viewed as test particles that act as probes of the various dynamical processes in the $\mathcal{N} = 4$ plasma background. This is because, in the large N_c limit, the influence of fundamental representation fields on bulk properties of the plasma is negligible.

The $\mathcal{N} = 4$ SYM is studied because it is easier than QCD, there hasn't been any techniques that provide a good approximation of dynamical processes in strongly coupled quantum field theories in real time. Rates for equilibrium, like the rate that a moving heavy quark losses its energy, can't be obtained directly from Euclidean correlation functions, as a result they are inaccessible in Monte Carlo lattice simulations [90]. However, the AdS/CFT conjecture states that the $\mathcal{N} = 4$ $SU(N_c)$ theory is exactly equivalent to type IIB string theory which is approximated by classical type IIB supergravity at large N_c and λ [187, 188]. This approximation allows calculations that cannot be performed perturbatively in QCD to be translated into calculations in classical general relativity.

Type IIB strings are represented by the string coupling, g_s as well as the tension on the string, T_0. This can be replaced by the fundamental string length scale given by $l_s \equiv (2\pi T_0)^{-1/2}$. The background is described by the radius of curvature of the AdS_5 space (L), which is required to be the same as the radius of S^5. The AdS/CFT correspondence equips us with a way to translate between these two theories. Table (C.1) gives a summary of the relationship of various parameters in the two theories.

In the $\mathcal{N} = 4$ SYM plasma, the static thermal mass of the quark $M_{rest}(T)$ also gives the free energy of a stationery quark and in the zero temperature limit, it is equal to the Lagrangian mass of the quark, m. In the context of classical general relativity, increasing the temperature of the gauge theory corresponds to the addition of a black hole in the centre of AdS_5 [191], meaning that the Hawking temperature of the black hole is equivalent to the temperature of the gauge theory. The zero temperature in QCD cannot be described by $\mathcal{N} = 4$ SYM since

Table C.1: Translation of various parameters in AdS/CFT [90].

AdS	$\mathcal{N} = 4$ SYM	Quantity
L	$-$	AdS^5 and S^5 curvature radius
l_s	$\lambda^{1/4} L$	fundamental string scale $[\equiv \sqrt{\alpha'}]$
$(L/l_s)^4$	λ	't Hooft coupling $[\equiv g_{YM}^2 N_c]$
T_0	$\sqrt{\lambda} L^{-2}/2\pi$	string tension $[(2\pi l_s^2)^{-1}]$
g_s	$g_{YM}^2/4\pi$	string coupling
u_h	πT	radius of a black hole horizon $(\times L^{-2})$
u_h/π	T	temperature
u_m	$\frac{2\pi}{\sqrt{\lambda}}(M_{rest} + \Delta m)$	minimal radius of D7-brane $(\times L^{-2})$
$T_0 L^2 u_h$	$\Delta m(t)$	thermal rest mass shift $[= \sqrt{\lambda} T/2]$
$T_0 L^2 (u_m - u_h)$	$M_{rest}(T)$	static thermal quark mass

the properties of the two are different, $\mathcal{N} = 4$ SYM is a conformal theory that doesn't have a particle spectrum, on the other hand, QCD has a particle interpretation [90]. However, at non-zero temperatures (as well as higher temperatures found in QCD), each of the theories describe hot, non-Abelian plasmas that have a qualitatively similar Debye screening, finite spatial correlation lengths as well as hydrodynamic behavior [26]. One big difference between the two theories is that all excitations (i.e gluons, fermions and scalars) in an $\mathcal{N} = 4$ SYM plasma are in the adjoint representation, on the other hand, a hot QCD plasma contains fundamental representation quarks and only gluons are in adjoint representation [90].

A number of reasons have been given to describe why many of the properties of non-Abelian plasmas that are strongly coupled may not be sensitive to what the plasma is composed of or the exact interaction strength. As $\lambda \to \infty$ in $\mathcal{N} = 4$ SYM, the limits of bulk thermodynamic quantities like the energy density, pressure, entropy density and transport coefficients like the shear viscosity, are finite. The ratio of the pressure to the free Stefan-Boltzmann limit (counts the number of degrees of freedom) in strongly coupled $\mathcal{N} = 4$ SYM is similar to the equivalent ratio in QCD at temperatures that are a few times T_c [24]. In $\mathcal{N} = 4$ SYM, $\eta/s = 1/4\pi$, the same value is obtained in all other theories that have gravity duals [192, 193] as $\lambda \to \infty$, even though this value is less than that found in weakly coupled theories or any known material [26], it is however in good agreement with hydrodynamic models at RHIC energies [194, 195].

Given that strongly coupled $\mathcal{N} = 4$ gauge theory is significantly different from QCD, the translation of parameters such as the drag loss coefficient (μ) and transport coefficients (κ_T and κ_L) from $\mathcal{N} = 4$ SYM into quantitative predictions in QCD is highly nontrivial. It introduces a considerable uncertainty in the problem which can be characterised by the following two comparison schemes [89, 196].

C.1.1 "Gubser" parameters

The results in Ref. [90, 140] showed that the drag force that is experienced by a heavy quark propagating through a thermal state of $SU(N_c)$ $\mathcal{N} = 4$ SYM theory is given by Eq. (C.2)

$$F_{drag} = -\frac{\pi\sqrt{g_{YM}^2 N_c}}{2}T^2\frac{v}{\sqrt{1-v^2}} \tag{C.2}$$

in the large N_c and large 't Hooft's coupling limit. The heavy quark's momentum falls by $1/e$ in a time given by Eq. (C.3),

$$t_D = \frac{1}{\mu} = \frac{2m}{\pi\sqrt{g_{YM}^2 N_c}T^2} \tag{C.3}$$

which is found from using $F_{drag} = dp/dt$ and $p/m = v/\sqrt{1-v^2}$. Using a different approach to the non-relativistic limit taken in Ref. [91], one finds the diffusion coefficient for heavy quarks given in Eq. (C.4).

$$D = \frac{2}{\pi T\sqrt{g_{YM}^2 N_c}} \tag{C.4}$$

As a result, we need to know what values to use for the 't Hooft's coupling $(g_{YM}^2 N_c)$ when comparing any of these results to experimental data. Several approaches have been attempted, such as taking $g_{YM} = g_s$, resulting in $\alpha_s = g_s^2/4\pi$, where g_s is the coupling in the QCD Lagrangian. However, this prescription results in the Debye mass (m_D) in $\mathcal{N} = 4$ SYM being substantially larger than that of QCD at weak coupling [197]. We'll now discuss the approach carried out in Ref. [196], where the 't Hooft coupling is normalised by comparing the force between a static quark and anti-quark to the predictions such as those found in lattice gauge theory, then compare parameters in $\mathcal{N} = 4$ SYM to those in QCD at fixed energy density instead of fixed temperature. Computing the force between a quark and anti-quark in string theory can be done using the Wilson loop construction in the AdS_5-Schwarzschild geometry, for the zero-temperature case, with the potential given by Eq. (C.5).

$$V(r) = -g_{YM}^2 N_c\frac{4\pi^2}{\Gamma(1/4)^4}\frac{1}{r} \tag{C.5}$$

In the non-zero temperature case, the radius and free energy are expressed parametrically as follows:

$$r = \frac{2}{\pi T}\tilde{r}(q), \quad F = T\sqrt{g_{YM}^2 N_c}\,\tilde{F}(q) \tag{C.6}$$

$$\tilde{r}(q) \equiv q\sqrt{1-q^4}\int_0^1 du \frac{u^2}{\sqrt{1-u^4}\sqrt{1-q^4u^4}} \tag{C.7}$$

$$\tilde{F}(q) \equiv \frac{1}{q}\left[\int_0^1 \frac{du}{u^2}\left(\sqrt{\frac{1-q^4u^4}{1-u^4}}-1\right)-1+q\right] \tag{C.8}$$

where r is the distance between the quarks, F is the free energy and q is a dimensionless parameter that describes how far down into AdS_5-Schwarzschild the string dangles. The dimensionless forms of the radius ($\tilde{r}$) and the free energy ($\tilde{F}$) are introduced for convenience. The integrals can be computed by making use of hypergeometric functions and one finds that F is negative only for $q < q_* \approx 0.66$ and the force between the quark and anti-quark drops abruptly to zero for $\tilde{r} > \tilde{r}_* \approx 0.38$ fm [140] as opposed to the exponential behavior observed for Debye screening. At plasma temperatures of $T = 250$ MeV and comparing QCD to $\mathcal{N} = 4$ SYM at the same temperature, the rescaled radius, $\tilde{r}_* = 0.19$ fm.

Lattice calculations for QCD with two [198] and three [199] flavours with a Debye radius (characterises the distance at which medium modifications in the quark anti-quark interaction become dominant) that is defined by inspecting the large behavior of the lattice free energy in the presence of a quark and an antiquark give $r_D = 0.24$ fm [140, 198] at $T = 250$ MeV (with $T_c = 190$ MeV). This result is close to r_*, however, quantities can also be compared at the radius where the zero-temperature potential equals the large distance limit of the free energy, $r_{med} \approx 0.42$ or at the radius where a quantity $\alpha_{q\bar{q}}(r)$ is maximised, $r_{max} \approx 0.33$ [198]. These results are discussed extensively in Ref. [198] and show that at temperatures $> \Lambda_{QCD}$, there are little to no differences between r_{max} and r_{med} obtained in either pure gauge ($N_f = 0$) and QCD with $N_f = 2, 3$. In addition, at high temperatures r_{med} drops similarly to the Debye screening radius which drops like $1/gT$ from perturbation theory.

Counting the degrees of freedom shows that there are three times more in $\mathcal{N} = 4$ SYM than in QCD and ultimately explains why for fixed temperatures, the screening length is smaller in $\mathcal{N} = 4$ SYM compared to QCD. If g_* is defined through Eq. (C.9),

$$\epsilon = g_*\frac{\pi^2}{30}T^4 \tag{C.9}$$

then in the weakly coupled limit of $\mathcal{N} = 4$ with gauge group SU(3), $g_* = 120$ while $g_* = 37$ and $g_* = 47.5$ for QCD with two flavours and three flavours respectively in the same limit [140]. On the other hand, $g_* \approx 90$ in strongly coupled $\mathcal{N} = 4$ SYM [24], while $g_* \approx 33$ for QCD with $2+1$

flavours in the same limit [200]. Given that the energy density is exactly proportional to T^4 ($\epsilon \propto T^4$) in $\mathcal{N} = 4$ SYM (due to exact conformal invariance) and is approximately proportional to T^4 in QCD for $T \gtrsim 1.1T_c$, then the condition given in Eq. (C.10)

$$T_{SYM} \approx 3^{-1/4}T_{QCD} \tag{C.10}$$

is required in order to have $\epsilon_{SYM} = \epsilon_{QCD}$ [140]. Since the quantity r_* doesn't depend on $g_{YM}^2 N_c$ in the large $g_{YM}^2 N_c$ limit, it cannot be used to normalise the 't Hooft coupling. An approach to normalise the 't Hooft coupling requires a comparison of the magnitude of the static force between the quark and anti-quark in string theory and in lattice QCD. The quantity of interest in lattice simulations is $\alpha_{q\bar{q}}(r)$ or $\alpha_{q\bar{q}}(r, T)$ for zero and non-zero temperature;

$$\alpha_{q\bar{q}}(r) = \frac{3}{4}r^2\frac{dV}{dr} \tag{C.11}$$

$$\alpha_{q\bar{q}}(r, T) = \frac{3}{4}r^2\frac{\partial F}{\partial r} \tag{C.12}$$

At zero temperature, the potential between the quark and anti-quark can be approximated by a Coulomb potential plus linear terms as shown in Eq. (C.13)

$$V(r) = -\frac{4\alpha/3}{r} + \sigma r, \quad \alpha = 0.212, \quad \sigma = (420 MeV)^2 = (0.47 fm)^{-2} \tag{C.13}$$

for $0.1 \gtrsim r \gtrsim 1.2$ fm and $\alpha_{q\bar{q}} \approx \alpha_s$ for $r < 0.1$ fm with two flavours. At non-zero temperature, Fig (C.1) shows $\alpha_{q\bar{q}}(r, T)$ obtained in lattice simulations of two-flavour QCD [198] compared to an analogous quantity in $\mathcal{N} = 4$ SYM given by Eq. (C.14).

$$\alpha_{SYM}(r, T_{SYM}) = \frac{3}{4}r^2\frac{dV}{dr} \tag{C.14}$$

The values used in Ref. [140] are as follows:

$$3.5 = [g_{YM}^2 N_c]_{lower} \lesssim g_{YM}^2 N_c \lesssim [g_{YM}^2 N_c]_{upper} = 8 \tag{C.15}$$

$$g_{YM}^2 N_c \sim [g_{YM}^2 N_c]_{typical} = 5.5 \tag{C.16}$$

where Eq. (C.15) is the representative value which gives the 't Hooft's coupling $\lambda = 5.5$ by equating $\alpha_{SYM}(r, T_{SYM})$ and $\alpha_{q\bar{q}}(r, T_{QCD})$ at approximately the largest r where $\alpha_{SYM}(r, T_{SYM})$ is defined, known to be $1/(\pi T_{QCD})$ for $T_{QCD} = 250$ MeV [140]. In Fig. (C.1), the radius r (fm) is in log scale and a)compares $T_{SYM} = 190$ MeV to $T_{QCD} = 250$ MeV while b)compares $T_{SYM} = T_{QCD}$. The thick black curve corresponds to α_{SYM} given in Eq. (C.14) for $g_{YM}^2 N_c = 5.5$, while the thin upper curve corresponds to $g_{YM}^2 N_c = 8$ and the thin lower curve to $g_{YM}^2 N_c = 3.5$. The dots show results from lattice simulations [198], where the red dots correspond to $T/T_c = 1.23$, the green dots to $T/T_c = 1.37$ and the blue dots to $T/T_c = 1.5$. The dashed grey curve is the result for $\alpha_{q\bar{q}}(r)$ at zero-temperature and it cuts the SYM curve

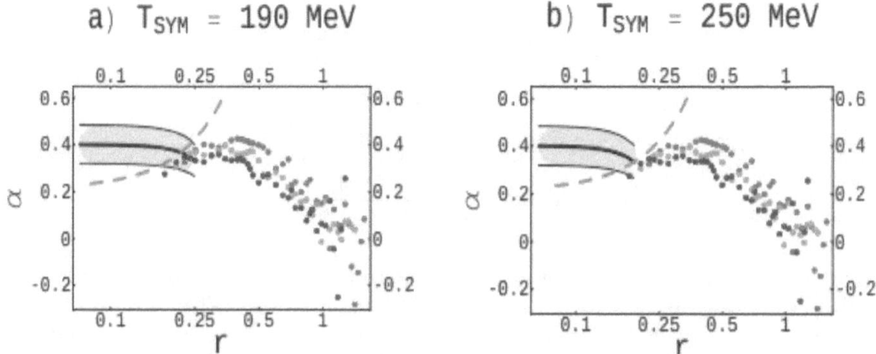

Figure C.1: A comparison between the static force of a quark and anti-quark for $\mathcal{N} = 4$ SYM and two flavour QCD for two different values of the plasma temperature T_{SYM}, obtained in Ref. [140]

at $r_* = 0.25$ fm for $T_{SYM} = 190$ MeV and $r_* = 0.19$ fm for $T_{SYM} = 250$ MeV [140].

In summary, energy loss calculations using the **Gubser (Gb)** set of parameters correspond to $\mathbf{T_{SYM} = 3^{-1/4}T_{QCD}}$, giving the relationship between temperature of the SYM plasma to that of the QCD plasma and the 't Hooft's coupling $\boldsymbol{\lambda = 5.5}$.

C.1.2   "Reasonable" parameters

In this scheme, the temperature of the SYM plasma is the same as the QCD plasma and the 't Hooft's coupling is given by Eq. (C.18) [88].

$$T_{SYM} = T_{QCD} \tag{C.17}$$
$$\lambda = 4\pi\alpha_s N_c = 4\pi \times 0.3 \times 3 \approx 11.3 \tag{C.18}$$

The strong coupling, α_s, taken to be 0.3 in this case, varies with the scale and its value is usually given at a specific reference scale, $Q^2 = M_Z^2$, where Q is the momentum transfer and M_z is the Z^0 mass. The running coupling is logarithmic with energy, governed by a 'beta function' as shown by the following equations; with a solution given by Eq. (C.22) [201].

$$Q^2 \frac{\partial \alpha_s}{\partial Q^2} = \frac{\partial \alpha_s}{\partial ln(Q^2)} = \beta(\alpha_s) \tag{C.19}$$

$$\beta(\alpha_s) = -\alpha_s^2(b_0 + b_1\alpha_s + b_2\alpha_s^2 + ...) \tag{C.20}$$

$$b_0 = \frac{11(3) - 2n_f}{12\pi}, \quad b_1 = \frac{153 - 19n_f}{24\pi^2} \tag{C.21}$$

$$\alpha_s(Q^2) = \alpha_s(M_Z^2)\frac{1}{1 + b_0\alpha_s(M_Z^2)ln\left(\frac{Q^2}{M_Z^2}\right) + O(\alpha_s^2)} \tag{C.22}$$

where the first term in the coefficient b_0 is due to gluon loops, the second term is due to quark loops and the coefficient b_1 accounts for double quark/gluon loops. The slope of α_s depends on the number of flavours, n_f [201] (see the β function) and effectively decreases continuously with increasing momentum transfer (i.e shorter distances) leading to asymptotic freedom as shown on the right of Fig (C.2). On the other hand, the coupling increases logarithmically with decreasing momentum transfer (i.e larger distances), however, this increase is limited by the finite size of the proton (colour confinement) [202].

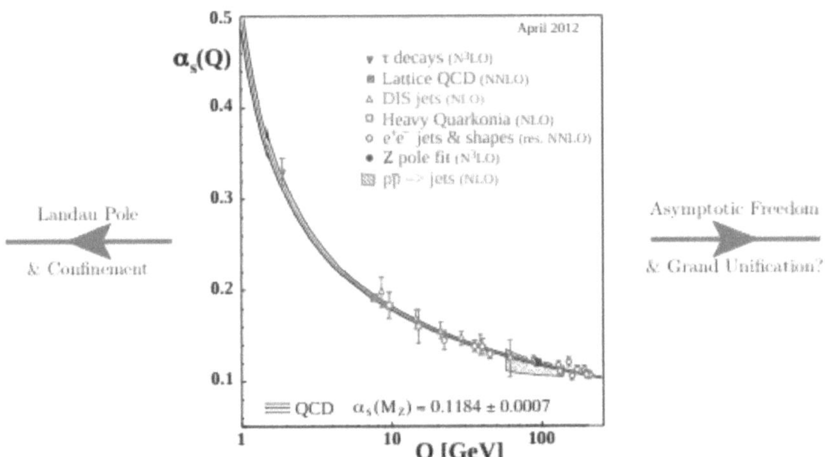

Figure C.2: The running strong coupling (α_s) with respect to the energy scale (Q) in a theoretical framework (band) as well as in physical processes at different characteristic scales [203]

C.2 Heavy quark speed limits

The speed limit of the heavy quark can be obtained by ensuring consistency on the derivation of the drag formulae, which requires the heavy quark to be moving at a constant velocity [88]. This is done by analysing single quark solutions for the relevant equations of motion for an open string in the AdS_5 black hole background. As discussed in Sec. (C.1), finite temperature $\mathcal{N} = 4$ $SU(N_c)$ SYM theory has a gravity dual obtained by S^5 times a five dimensional AdS-black hole solution [188]. The resulting geometry consists of a black hole containing a flat four dimensional horizon sitting inside the AdS space with the following metric in $d+1$ dimensions;

$$ds^2 \;=\; L^2 \left(\frac{du^2}{h(u)} - h(u)dt^2 + u^2 \delta_{ij} dx^i dx^j \right) \tag{C.23}$$

$$h(u) \;=\; u^2 \left[1 - \left(\frac{u_h}{u} \right)^d \right] \tag{C.24}$$

where u is the radial coordinate, rescaled by L^{-2} and $h(u)$ vanishes at the black hole horizon where $u = u_h$. The radius of the black hole horizon is related to the Hawking temperature (which is equal the temperature of the field theory dual) by;

$$T \;=\; \frac{d}{4\pi} u_h, \quad u_h = \pi T, \quad d = 4 \tag{C.25}$$

Introducing a flavour of fundamental representation quarks is equivalent to adding of a D7-brane [204] which wraps an S^3 inside the transverse S^5 and fills the entire asymptotically AdS space down to the lowest radius $u = u_m > u_h$. If we have an open string ending on the D7-brane, the quark is described by the string's endpoint and u_m gives the mass of the quark. The string equation of motion given in Eq. (C.28) is derived from the Nambu-Goto action given by Eq. (C.26)

$$S \;=\; -T_0 \int d\sigma d\tau \sqrt{-\det g_{ab}} \tag{C.26}$$

$$g_{ab} \;=\; \begin{bmatrix} -h + u^2 (\dot{x})^2 & u^2 \dot{x} x' \\ u^2 \dot{x} x' & \frac{1}{h} + u^2 (\dot{x})^2 \end{bmatrix} \tag{C.27}$$

$$0 \;=\; \frac{\partial}{\partial u} \left(h u^2 \frac{x'}{\sqrt{-g}} \right) - \frac{u^2}{h} \frac{\partial}{\partial t} \left(\frac{\dot{x}}{\sqrt{-g}} \right) \tag{C.28}$$

The corresponding canonical momentum densities are

$$\begin{pmatrix} \pi_x^0 \\ \pi_u^0 \\ \pi_t^0 \end{pmatrix} = \frac{T_0 L^4}{\sqrt{-g}} \begin{bmatrix} \dot{x} u^2 h^{-1} \\ -\dot{x} x' u^2 h^{-1} \\ -1 - (x')^2 u^2 h^1 \end{bmatrix}, \quad \begin{pmatrix} \pi_x^1 \\ \pi_u^1 \\ \pi_t^1 \end{pmatrix} = \frac{T_0 L^4}{\sqrt{-g}} \begin{bmatrix} -x' u^2 h^1 \\ -1 + (\dot{x})^2 u^2 h^{-1} \\ (\dot{x})^2 x' u^2 h^1 \end{bmatrix}, \tag{C.29}$$

where π_t^0 gives the energy density and π_x^0 gives the x-component of the momentum density on the string worldsheet. Their respective integrals along the string give the corresponding total energy and total momentum of the string.

$$E \;=\; \int d\sigma \pi_t^0, \quad p = \int d\sigma \pi_x^0 \tag{C.30}$$

The simplest solution to the string equations of motion is just a constant, $x(u,t) = x_0$, describing a stationery string that stretches from $u = u_m$ to $u = u_h$ and it describes a stationery quark that is resting in the thermal medium [90]. The total momentum and momentum density vanish and the energy is given by;

$$E \;=\; T_0 L^2 \int_{u_h}^{u_m} du = T_0 L^2 (u_m - u_h) \tag{C.31}$$

and is equal to the quark's Lagrangian mass in the zero temperature limit, $T_0 L^2 u_m = m$. If u_m is increased (i.e move the D7-brane to a bigger radius), that would correspond to increasing the mass of the quark.

Another solution to the string equation of motion is a rigidly moving string profile, $x(u,t) = x_0 + vt$. This solution is unphysical since $-g$ is not positive definite in this case [90] and is therefore discarded. Taking $x = vt$ gives inconsistent initial conditions such that at $t = 0$, some parts of the string are moving at a velocity that is greater than the local speed of light. Stationary solutions of the form $x(u,t) = x(u) + vt$ are used to find a physical configuration corresponding to a quark that is moving with constant velocity. The equation of motion becomes;

$$\frac{d}{du}\left(hu^2 \frac{x'}{\sqrt{-g}} \right) \;=\; 0 \tag{C.32}$$

integrating once gives,

$$(x')^2 \;=\; \frac{C^2 v^2}{u^8}[1 - (u_h/u)^d]^{-2} \frac{1 - v^2 - (u_h/u)^d}{1 - C^2 v^2 u^{-4} - (u_h/u)^d} \tag{C.33}$$

where C is an integration constant that determines the momentum current flowing along the string, with π_t^1 and π_x^1 remaining constant along the string. This gives

$$\frac{-g}{L^4} \;=\; \frac{1 - v^2 - (u_h/u)^d}{1 - C^2 v^2 u^{-4} - (u_h/u)^d} \tag{C.34}$$

where the numerator and denominator are both positive if u is large and negative for u close to u_h and $-g$ can only remain positive everywhere on the string if these change sign at the same point. This fixes the constant C to;

$$C \;=\; \pm\left(\frac{u_h^d}{1 - v^2} \right)^{2/d} = \pm\frac{u_h^2}{\sqrt{1 - v^2}}, \quad d = 4 \tag{C.35}$$

this yields

$$x'(u) = \pm v \frac{u_h^2}{h(u)u^2} \tag{C.36}$$

$$x(u,t) = x_{\pm}(u,t) \equiv x_0 \pm vF(u,v^2) + vt \tag{C.37}$$

$$F(u) = \frac{1}{2u_h}\left[\frac{\pi}{2} - \arctan\left(\frac{u}{u_h}\right) - \operatorname{arccoth}\left(\frac{u}{u_h}\right)\right], \quad d = 4 \tag{C.38}$$

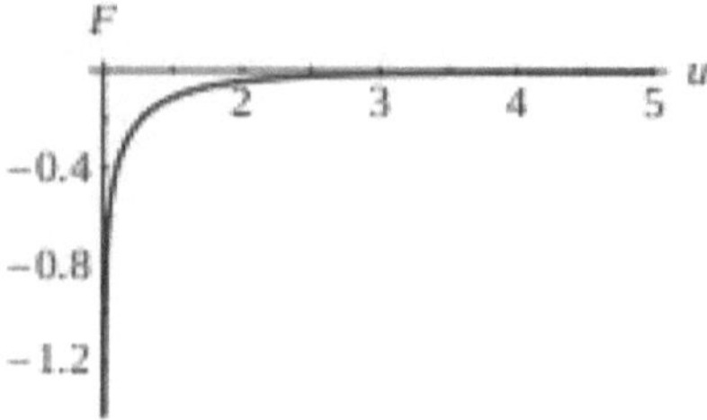

Figure C.3: The function F(u) which determines the string profile, in units $u_h = 1$ obtained in Ref. [90]

The function $F(u)$ is given in Fig (C.3) for $d = 4$, it vanishes in the limit $u \to \infty$ and diverges in the limit $u \to u_h$. The energy flux, $\pi_t^1 = T_0 L^2 C v^2$, is proportional to C and if $C > 0$, energy will flow down the string towards the horizon, while if $C < 0$, energy flows upward from the horizon towards the moving quark. The physical solution corresponds to the case $C > 0$ with outgoing (from the physical region into the black hole) boundary conditions at the horizon [90, 205]. The rates at which the energy and momentum flow toward the horizon are given by;

$$\pi_t^1|_{u=u_h} = T_0 L^2 u_h^2 \frac{v^2}{(1-v^2)^{2/d}} = \frac{\pi}{2}\sqrt{\lambda}T^2\frac{v^2}{\sqrt{1-v^2}}, \quad d = 4 \tag{C.39}$$

$$-\pi_x^1|_{u=u_h} = T_0 L^2 u_h^2 \frac{v}{(1-v^2)^{2/d}} = \frac{\pi}{2}\sqrt{\lambda}T^2\frac{v}{\sqrt{1-v^2}}, \quad d = 4 \tag{C.40}$$

The stationary solution $x_+(u,t)$ describes an open string running from the AdS boundary ($u = \infty$) and asymptotically approaching the horizon at $u = u_h$. It can be regarded as an open string that runs from a D7-brane with minimal radius u_m down to the horizon by truncating the solution $x_+(u,t)$ at an arbitrary radius $u_m > u_h$. The solution with a constant $\pi_x^1 \neq 0$ does not satisfy the standard Neumann boundary conditions which require the momentum flux, π_x^1 to vanish at the flavour brane. The solution is physical, but requires a force to act on the string endpoint and feed energy and momentum into the string [90]. The force is provided by a constant electric field on the flavour brane.

If a quark is moving with velocity v, in the presence of a constant external electric field, its equation of motion is [109]

$$\frac{dp}{dt} = -\mu p + E \tag{C.41}$$

where $\mu \sim \sqrt{\lambda} T^2 / M_Q$ and E is the electric field whose energy can be increased to accelerate the quark to its terminal velocity and balances the energy and momentum loss of the heavy quark. The electric field on a D-brane can only be increased to a critical value given in Eq. (C.43) and computed from the Born-Infeld action for the probe brane in the AdS geometry given in Eq. (C.42) [109].

$$S_{BI} \sim \sqrt{1 - \left(2\pi\alpha' \frac{R^2}{r^2}\right)^2} \tag{C.42}$$

$$E < \frac{M^2}{\sqrt{\lambda}} \tag{C.43}$$

Beyond the critical value of E, the force pulling the endpoints of the string apart overcomes the tension and the system becomes unstable. The speed limit is then obtained as follows;

$$E \sim \mu p = \frac{M^2}{\sqrt{\lambda}} \tag{C.44}$$

$$\implies \gamma < \left(\frac{M}{\sqrt{\lambda} T}\right)^2 \tag{C.45}$$

In this description, the diffusion and drag coefficients are related by the Einstein relation as shown in Eq. (C.46) while the diffusion constant, D, is related to the diffusion coefficient by Eq. (C.47)

$$\mu = \frac{\kappa}{2MT} \tag{C.46}$$

$$D = \frac{T}{M\mu} = \frac{2T^2}{\kappa} = \frac{2}{\pi\sqrt{\lambda} T} \tag{C.47}$$

As a result, the momentum fluctuations are given by Eq. (C.48) [91], while the jet quenching parameter $\hat{q}$ is given by Eq. (C.49) [139]. Computations using these parameters will be labelled **D=const** since the diffusion coefficient does not depend on the heavy quark's momentum and the momentum fluctuations obey the fluctuation-dissipation theorem.

$$\kappa = \pi\sqrt{\lambda} T^3 \tag{C.48}$$

$$\hat{q} = <p_\perp(t)^2> \lambda \approx \kappa t / \lambda = (2\pi T^3 \sqrt{\lambda})/v \tag{C.49}$$

Appendix D

Collective Flow

Collective flow describes the correlated emission of produced particles and provides a phenomenological description of a collective expansion and the azimuthal distribution of produced particles. The distribution of particles in the azimuthal angle (ϕ) of the detected particle in the transverse plane is given by a Fourier expansion as shown by Eq. (D.1),

$$E\frac{d^3N}{d^3p} \;=\; \frac{1}{2\pi}\frac{d^2N}{p_T dp_T dy}\left(1 + 2\sum_{n=1}^{\infty} v_n(p_T, y)\cos[n(\phi - \Psi_R)]\right) \tag{D.1}$$

where E is the energy of the particle and y is the rapidity. The differential flow coefficients given by Eq. (D.2) [206],

$$v_n(p_T, y) \;=\; \langle\cos[n(\phi - \Psi_R)]\rangle = \frac{\int d\phi \frac{d^3N}{p_T dp_T dy}\cos[n(\phi - \Psi_R)]}{\int d\phi \frac{d^3N}{p_T dp_T dy}} \tag{D.2}$$

The expansion of the momentum integrated invariant distribution of detected particles is given in Eq. (5.11) and the corresponding flow coefficients are given by Eq. (5.12). In Fig. (D.1) we show two different configurations of a collision, one where the participating plane angle coincides with the reaction plane and where it does not. The angle Ψ_{PP} is the plane of the symmetry of the initial collision zone and it coincides with the reaction plane for smooth initial matter distribution as shown in Fig. (D.1 a). The azimuthal angle (ϕ) should always be measured with respect to the angle of the participating plane instead of the angle of the reaction plane, unless the two coincide. Determining the position of the reaction plane ($\Psi_R = \Psi_{RP}$) is discussed in Ref. [207].

Note that the Fourier expansion doesn't include sine terms since they cancel due to the reflection symmetry with respect to the reaction plane. The various flow coefficients allow us to classify the collective flow into three main types as shown in Fig. (D.2). The coefficient v_1 is the directed flow, while v_2 is the elliptic flow and v_3 is the triangular flow. The angle of the reaction plane cannot be measured directly in relativistic heavy-ion collisions, it is estimated

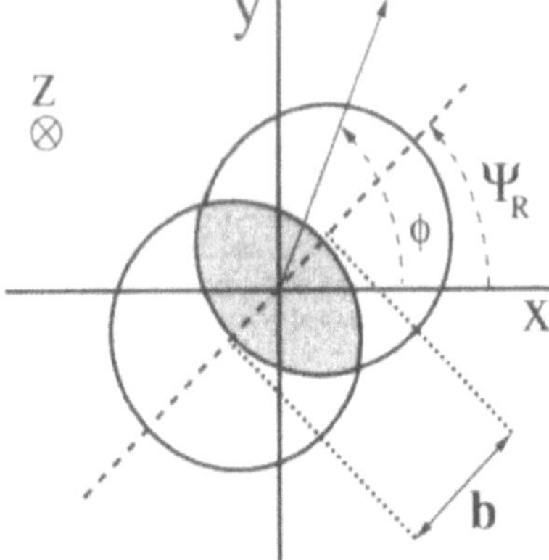
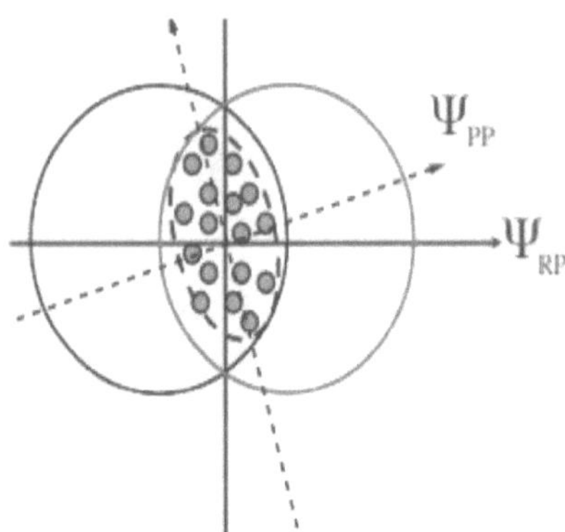

(a) Coordinate system with the reaction plane coinciding with symmetry plane obtained in Ref. [206].

(b) Simulation of participating nucleons in the transverse plane using the Monte Carlo Glauber approach and showing an overlap region that is tilted with respect to the reaction plane obtained in Ref. [68].

Figure D.1: Coordinate systems showing the two cases where the participating plane angle coincides with the reaction plane and where it does not.

from the measured particle azimuthal distribution on an event-by-event basis.

Radial flow occurs only in central collisions of spherical nuclei and has no defined reaction plane. It has an isotropic expansion in the transverse plane (plane perpendicular to the collision axis). The transverse spectra of the produced particles are the combined result of thermal emission and radial flow. This is described by the blast wave model discussed in Ref. [208].

Directed flow is characterised by having a direction and $v_1 \neq 0$ and has opposite signs in the two hemispheres. It comes from momentum conservation, in that, the flow of particles originating from one of the nuclei must be counterbalanced (equal magnitude but opposite direction), to the flow of particles from the other nuclei.

Elliptic flow is the azimuthal momentum anisotropy that occurs in non-central ($b > 0$) heavy-ion collisions. Studies of the elliptic flow are very crucial in building our understanding of thermalisation, in that, the elliptic flow signals the presence of interactions between the constituents of the QGP. We expect more interactions to lead to a larger elliptic flow (in terms of magnitude) and the larger the magnitude, the quicker the system thermalises. As a result, the

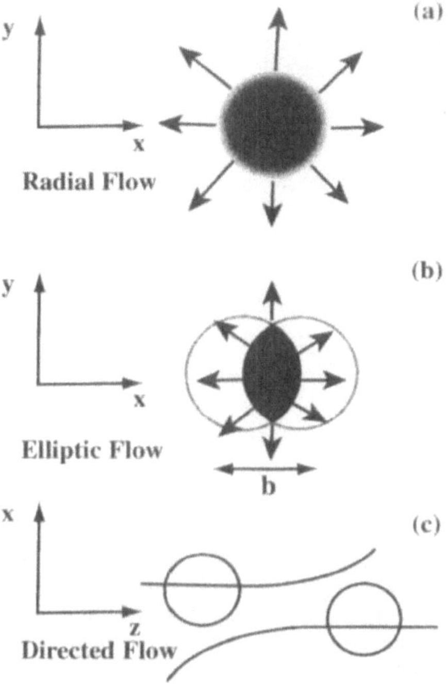

Figure D.2: Three types of flow phenomena from heavy-ion collisions, obtained in Ref. [206].

magnitude of the elliptic flow is also a direct probe of the level of thermalisation in the system.

The elliptic flow depends on the collision impact parameter, and in practice, it needs to be analysed for narrow regions of centrality. It also has a transverse momentum dependence and peaks at $p_T \approx 3$ GeV/c, as well as a mass dependence such that heavier particles are more affected by the flow [209]. In general, the elliptic flow is largest in mid-central collisions where the impact parameter is of the order of the nuclear radius and drops in peripheral collisions from a lack of collective flow [152, 210].

The observation of a large elliptic flow at RHIC [211] was one of the discoveries that paved our understanding that QGP interacts strongly. Anisotropic flow is very sensitive to the early particle interactions and develops shortly after the collision since spatial asymmetries rapidly decrease with time. Therefore, it is an observable that gives us a direct probe to the QGP medium. Various explanations have been given for anisotropic flow, such as constituent re-scattering and there are discussions on other possibilities of the origin of elliptic flow such as the partonic structure of the nuclei [212], colour dipole orientation [213] and possibly the direct anisotropy in particle emission from the Color Glass Condensate (CGC) [214, 215].

An alternative approach to compute the flow coefficients in relativistic nuclear collisions is suggested in Ref. [216]. The azimuthal distribution function of the quantity being considered (i.e $dN/d\phi$) is given by $f(\phi)$ and taking its Fourier expansion as given by Eq. (D.3 - D.5);

$$f(\phi) \;=\; \frac{a_0}{2\pi} + \frac{1}{\pi} \sum_{n=1}^{N} \left(a_n \cos(n\phi) + b_n \sin(n\phi) \right) \tag{D.3}$$

$$a_n \;=\; \int_0^{2\pi} f(\phi) \cos(n\phi) d\phi \tag{D.4}$$

$$b_n \;=\; \int_0^{2\pi} f(\phi) \sin(n\phi) d\phi \tag{D.5}$$

the flow coefficients are then obtained from

$$v_n \;=\; \sqrt{a_n^2 + b_n^2} \tag{D.6}$$

If $v_n = 0$, $\forall n$, it implies that we have isotropic flow and if $v_n \neq 0$ then we have anisotropic flow. Odd Fourier harmonics have opposite sign in the forward and backward hemispheres, so only the even harmonics contribute. A hydrodynamics approach discussed in Ref. [166] is one of the models that are used to describe flow, however, it is only applicable in instances where the mean free path of the particles is much smaller than the system size. It describes the system using macroscopic quantities, this gives us a sense of the system's equation of state and the value of the speed of sound in the medium [68].

Experimentally, there is an uncertainty in the flow coefficients which is introduced by recon-structing the flow coefficients from many particle azimuthal correlations as these correlations have other contributions (non-flow) apart from anisotropic flow. Anisotropic flow can fluctuate from one event to the next (at fixed impact parameter). These flow fluctuations are introduced by various factors, such as fluctuations in the initial geometry of the overlapping region as the nuclei collide since the interaction between constituents of the colliding nuclei is random [217].

The left panel of Fig. (D.3) shows the density of binary collisions (shown in the overlap region of the colliding nuclei) in the transverse plane for $Au + Au$ collisions with $b = 7$ fm at $\sqrt{s} = 130$ GeV. The lines inside the overlap region are lines of constant density at 5%, 15%, 25%, . . .% of the maximum value. The dashed lines show the circumferences of the colliding nuclei (given by the Woods-Saxon distribution) as discussed in Chap. (2). The spatial asymmetry of the overlap region can be quantified by the spatial eccentricity defined by Eq. (D.7).

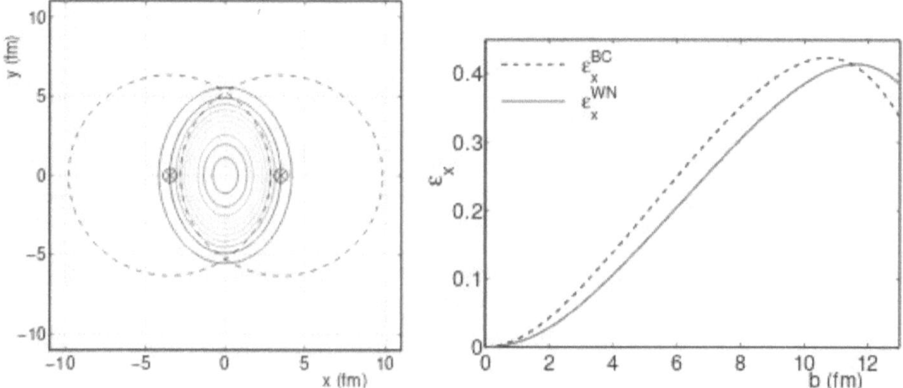

Figure D.3: (left) Binary collision density in the transverse plane for $Au+Au$ collision with $b = 7$ fm at $\sqrt{s} = 130$ GeV. (right) The geometric eccentricity with respect to the impact parameter for wounded nucleon and binary collision distributions for the same parameters [166].

$$\mathcal{E}_x = \frac{\langle y^2 - x^2 \rangle}{\langle y^2 + x^2 \rangle} \tag{D.7}$$

where x, y are the coordinates of the participating nucleons in the transverse plane and $\langle ... \rangle$ indicate averages taken with respect to the underlying density of binary nucleon-nucleon collisions (n_{BC} or n_{WN}, depending on the parameterisation used). The right panel of Fig. (D.3) shows the geometric eccentricity with respect to the impact parameter. In non-central collisions, the initial eccentricity is non-zero and positive as obtained for the two different underlying densities.

The transverse energy density profile of a non-central heavy-ion collision is shown in Fig. (D.4) as a function of time, contours indicate regions of constant energy density at 5%, $15 - 95\%$ of the maximum value. The black solid line indicates the transition into a mixed-phase, dashed line indicates the transition to a resonance gas phase and dashed-dotted line indicates the transition to the decoupled stage, where applicable. We can see how the energy density evolves into an almost symmetric system. The initial geometric anisotropy vanishes quickly and the momentum space distribution changes from approximately azimuthally symmetric to having a preferred direction in the reaction plane. This is due to the hydrodynamic evolution of the system, which is driven by its internal pressure gradients, this results in it expanding more strongly in the direction of the impact parameter since it is the shorter direction compared to

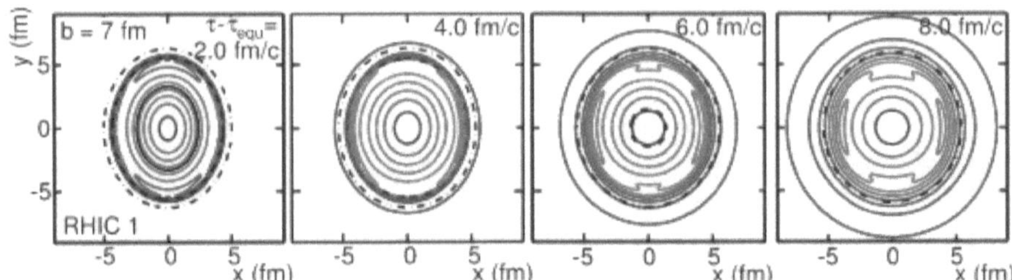

Figure D.4: The transverse energy density profile created in a non-central collision as a function of time with contours of constant energy density in the transverse plane for $Au + Au$ collision with $b = 7$ fm at $\sqrt{s} = 130$ GeV [166, 218, 219].

the direction perpendicular to the reaction plane with a smaller pressure gradient [166, 220].

The asymmetry in momentum space (momentum anisotropy) is quantified by Eq. (D.8),

$$\mathcal{E}_p \;=\; \frac{\langle T_{xx} - T_{yy}\rangle}{\langle T_{xx} + T_{yy}\rangle} = \frac{\int dx dy [T_{xx} - T_{yy}]}{\int dx dy [T_{xx} - T_{yy}]} \tag{D.8}$$

where T_{xx} and T_{yy} represents the diagonal transverse components of the energy momentum tensor and $\langle ... \rangle$ indicates an average over the transverse plane. The contour lines in Fig. (D.4) and their evolution with respect to time can be characterised quantitatively by the spatial eccentricity $\mathcal{E}_x(\tau)$. The sign conventions used here give a positive spatial eccentricity for out-of-plane elongation and gives a positive momentum anisotropy if the preferred direction of flow is towards the reaction plane.

The value of $\mathcal{E}_p$ starts at zero after the collision, $\mathcal{E}_p(\tau_{equ}) = 0$, as shown in Fig. (D.5). This is because the fluid starts from rest in the transverse plane, but as the fluid evolves, the anisotropy quickly develops due to re-scattering of particles. The solid lines have initial conditions achievable at RHIC with a realistic equation of state, while the dashed lines correspond to a massless ideal gas equation of state that uses a much higher initial energy density (with the initial temperature at the centre of the fireball, $T = 2$ GeV) [221]. It is expected to saturate at a certain time, $\tau \approx 6$ fm/c, this can be seen in Fig. (D.5) and occurs when the reaction zone attains azimuthal symmetry. This implies that elliptic flow is a self quenching phenomena, it is driven by the reaction zone asymmetry which continuously drops as the flow grows. The initial spatial asymmetry at the impact parameter used here ($b = 7$ fm) is $\mathcal{E}_x(\tau_{equ}) = 0.27$, but it goes to zero before the fireball matter reaches freeze out. This happens quicker in the case where the initial temperature is very high (dashed lines) and the source switches orientation at $\tau \approx 6$ fm/c and becomes elongated in-plane at late times [222].

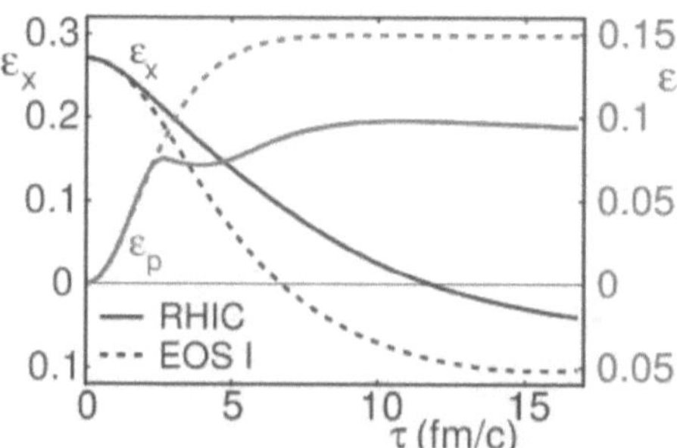

Figure D.5: Simulations of the temporal evolution spatial eccentricity ($\mathcal{E}_x$) as well as the momentum anisotropy ($\mathcal{E}_p$) as a function of time for $Au + Au$ collisions at RHIC with $b = 7$ fm at $\sqrt{s} = 130$ GeV using an ideal hydrodynamic model [166, 221].

The equation of state becomes very soft near a phase transition (i.e first order transition), this in turn inhibits transverse flow from being generated. As can be seen from the solid $\mathcal{E}_p$ curve in Fig. (D.5), this also affects the generation of transverse flow anisotropies. As more and more of the fireball enters the mixed phase, we see that $\mathcal{E}_p$, stops rising sharply and it decreases briefly implying that the system becomes more isotropic in both coordinate and momentum space. The final slight further increase in the momentum anisotropy occurs when the pressure gradients reappear after the phase transition is complete. The softness of the equation of state close to the phase transition means anisotropic flow is generated during earlier times in the system (i.e when the system is still entirely partonic, before hadronisation starts) [166].